MEDICAL PROCEDURES, TESTING AND TECHNOLOGY

ELISA

ADVANCES IN RESEARCH AND APPLICATIONS

MEDICAL PROCEDURES, TESTING AND TECHNOLOGY

Additional books and e-books in this series can be found on Nova's website under the Series tab.

MEDICAL PROCEDURES, TESTING AND TECHNOLOGY

ELISA

ADVANCES IN RESEARCH AND APPLICATIONS

JOSE MA. M. ANGELES
EDITOR

nova
Medicine & Health
New York

DOI: https://doi.org/10.52305/NIRE6124

NOTICE TO THE READER

Library of Congress Cataloging-in-Publication Data

ISBN: 978-1-68507-500-2

Published by Nova Science Publishers, Inc. † New York

Contents

PREFACE

Jose Ma. M. Angeles[*], *PhD*
Department of Parasitology, College of Public Health,
University of the Philippines Manila, Manila, Philippines

In September 1971, Swiss scientists Eva Engvall and Peter Perlmann published a communication in the journal Immunochemistry on enzyme-linked immunosorbent assay (ELISA) as a quantitative assay of immunoglobulin G. They have modified radioimmunoassay (RIA) by replacing the radioactive iodine with enzymes and used in determining the levels of IgG in rabbit serum.

Fifty years after, ELISA is still the most commonly used technique in measuring peptides, proteins, antibodies, antigens, and hormones in various samples. This plate-based assay is a very versatile laboratory technique and offers many advantages to laboratory scientists and researchers such as generating quick results, simple protocols, high sensitivity, being easily automated and being convenient to perform. Its application in the diagnosis of different infectious diseases has shown the unquestionable importance of this technique even in the modern world.

[*] Corresponding Author's E-mail: jmangeles3@up.edu.ph.

In this book, the usefulness of ELISA in the control of different parasitic diseases including schistosomiasis, babesiosis, and malaria will be discussed. There is also a chapter featuring an extensive discussion on the use of ELISA in functional food development and regulation. And finally, the last two chapters will present advanced techniques inspired by ELISA including a rapid immunoassay for detecting paralytic shellfish poisoning toxins and the enzyme-linked aptamer assay.

May this book provide significant information regarding the current advances and applications of ELISA and possibly inspire its use in another 50 years.

In: ELISA
Editor: Jose Ma. M. Angeles

ISBN: 978-1-68507-500-2

Chapter 1

UPDATES IN THE DIAGNOSTIC USE OF ELISA FOR SCHISTOSOMIASIS

Kevin Austin L. Ona*[1] *and Jose Ma. M. Angeles*[2,*]*, PhD
[1]College of Medicine, University of the Philippines Manila, Manila, Philippines
[2]Department of Parasitology, College of Public Health, University of the Philippines Manila, Manila, Philippines

ABSTRACT

Schistosomiasis, a parasitic infection caused by Schistosoma spp., remains to be a global public health concern. Central to its control and elimination is the prompt diagnosis of the disease. The value of serological diagnostics for schistosomiasis has been recently highlighted specifically in areas with low endemicity and almost nearing elimination where majority of individuals have light intensity of infection and might not be detected through stool microscopy. Enzyme-linked immunosorbent assay (ELISA) is one of the most commonly used serological methods in parasite diagnostics. Together with the identification and use of schistosome antigens, ELISA has become

* Corresponding Author's E-mail: jmangeles3@up.edu.ph.

widely used in schistosome diagnostics. However, certain drawbacks are still seen in immunoassay. Thus, developments and researches in ELISA and schistosome antigens are still ongoing with a focus on improving the accuracy, sensitivity, and specificity of the diagnostic technique. This chapter looks into some updates and advances in the diagnostic use of ELISA for schistosomiasis.

Keywords: schistosomiasis, diagnostics, ELISA, antigens

INTRODUCTION

Schistosomiasis is a parasitic disease caused by an infection with the zoonotic blood fluke *Schistosoma* spp. (WHO, 2018). The disease may present as an acute infection which may progress to a late chronic infection more commonly seen in patients residing in endemic regions (Gordon et al., 2019; McManus et al., 2018). Considered as one of the neglected tropical diseases, it remains to be a public health concern in 76 countries including China, Indonesia, and the Philippines where control of the parasite's transmission is heavily affected by the parasite's wide variety of definitive hosts and reservoirs (Berqguist and Gray, 2019; Blas et al., 2004; Gordon et al., 2019; Olveda and Gray, 2019; Wang et al., 2012). Aside from humans, schistosome's definitive hosts include common livestock like water buffalos, pigs, and goats, most of which are central to a community's livelihood (Berqguist and Gray, 2019; Francisco et al., 2019; Gordon et al., 2019; Rudge et al., 2008; Tianping and Jiaojiao, 2005; Zhou, Chen and Wu, 2005).

Diagnosis plays a key role in the control and management of schistosomiasis. Using the right diagnostic tools, preferably with high sensitivity and specificity, the disease may be identified and treated during its early stages. Diagnostics for schistosomiasis may either be parasitology–based or serology–based (Chen et al., 2021). Coproparasitology techniques, which include the Kato – Katz test, formalin – ether concentration technique (FECT), and FLOTAC, remain to be the most popular diagnostic choice (Utzinger et al., 2015; Weerakon et al., 2015). These methods, however, are time–consuming, have low sensitivity, and may yield false-negative results especially in regions of low prevalence, and in individuals with low infection intensity (Chen et al., 2021; Espiritu – Santo et al., 2014; Weerakon et al., 2015). For these reasons, some institutions have shifted towards using serology–based techniques are known for their higher sensitivity and specificity, and more time-efficient preparation and execution (Wang et al., 2012; Weerakon et al., 2015).

Serology–based diagnosis makes use of an immunodiagnostic approach where either direct or indirect detection of molecules derived from the host's immune response against the parasite is done (Doenhoff, Chiodini and Hamilton, 2004; Weerakon et al., 2015). Direct immunodiagnostics involves finding circulating antigens in the blood while indirect immunodiagnostics checks for the presence of antibodies against the said antigens (Doenhoff, Chiodini and Hamilton, 2004). It utilizes existing immunoassays, common of which is the enzyme-linked immunosorbent assay or ELISA. Immunodiagnostics is more time-efficient and recommended for point–of–care diagnosis and allows accurate detection of the parasite in cases of light infection intensity (Weerakon et al., 2015). The immunodiagnostic approach, however, has its disadvantages. In particular, the indirect or antibody detection techniques has low specificity; it may fail to discriminate between active and past infections and has been shown to display cross-reactivity against other helminths (Cai et al., 2017; Chen et al., 2021; Doenhoff, Chiodini and Hamilton, 2004). Alternatively, DNA – based diagnosis may be used since it offers higher sensitivity and specificity, and better accuracy as it exhibits low to no cross-reactivity (Cai et al., 2017; Cai et al., 2019; Chen et al., 2021; Weerakon, Gordon and McManus, 2018). Nucleic acid detection techniques, however, are costly and not recommended for large–scale application.

The growing interest in ELISA's use for schistosome diagnostics over the past few years has led to the discovery of more antigen candidates and further improvements. Here, we look into the updates in the use of ELISA as a diagnostic tool for Schistosomiasis.

Background

ELISA as Diagnostic Tool

ELISA uses an antibody with enzyme – conjugate to detect the presence of a ligand through specific binding. A substrate for the enzyme will then be introduced leading to a color change or production of a fluorescent signal. The color change or signal can be measured either qualitatively or quantitatively depending on the purpose of the assay. This immunoassay has long been used as an immunodiagnostic tool for several parasitic infections. It was initially assessed for schistosome antibody detection by Huldt et al., in 1975 where they saw its potential for diagnosing suspected patients and identifying the presence of infection in patients who previously tested negative for traditional diagnostic approaches (Huldt et al., 1975). When it comes to diagnosing schistosomiasis, ELISA is considered to be one of the most sensitive tools available (Sarhan et al., 2014; Turner et al., 2004; Zhu et al., 2010). It is more sensitive than the existing coproparasitology methods and is recommended for settings with low transmission of the parasite (Cesari et al., 2005). Some drawbacks, however, include its relatively low specificity and its difficulty in discriminating between current and previous infections (Ongongo, Kariuki and Wilson, 2018).

When compared with indirect hemagglutination assay (IHA), it is thought to be more sensitive, but less accurate. In an integrated meta-analysis study done by Wang, et al., on the *Schistosoma japonicum* immunodiagnostic efficacy between ELISA and IHA, ELISA was found to have higher sensitivity (57.1% to 97.3%) but lower specificity (20.4% to 84.3%) compared to IHA (37.6% to 95.1% and 35.7% to 93.8%, respectively). However, IHA's diagnostic efficacy exceeded that of ELISA in terms of diagnostic odds ratio which showed 9.41 (95% *CI*: 4.88 – 18.18) and 4.78 (95% *CI*: 3.21 – 7.13), respectively (Wang et al., 2012). This has been supported by Rahman et al., in their meta-analysis study on the accuracy of diagnostic tests where IHA with a diagnostic odds ratio (DOR) of 60 edged ELISA (DOR: 9.0) and PCR (DOR: 8.0) when it comes to detecting schistosomiasis in endemic areas in Asia (Rahman et al., 2021). Moreover, in a study by El – kady et al., among Egyptians infected by *Schistosoma haematobium*, the combination of ELISA and IHA gave a higher specificity (95%) than that of IHA alone (85%) which was equal to that of ELISA alone. The combination also had better sensitivity (100%) compared to IHA and ELISA alone (86% and 97%, respectively) (El – kady et al., 2018).

Interest in ELISA's use in schistosome diagnosis has increasingly grown since the 1970s to complement the limitations of coproparasitology techniques. This led to several improvements and developments in the assay and the discovery and exploration of antigens that we see today. In immunodiagnostics, ELISA is more commonly used for antibody detection where crude or purified schistosome antigens are used to capture antibodies in the sample serum (Ajibola et al., 2018; Cai et al., 2019; Weerakon et al., 2015). During infection, the body is exposed to a multitude of antigens as the parasite goes through its life cycle. Some of these are highly immunogenic and can elicit the production of a huge number of antibodies. These antigens (see table 1 and the end of the chapter for the summary) may then be used to target antibodies in the patient's serum. Some of the more commonly used antigens include the crude antigen preparations soluble egg antigen (SEA), adult worm antigen preparation (AWA), cercarial antigen (CA), and schistosomulum crude antigen (SCA), and the purified or recombinant single or multi-epitope antigens (Chen et al., 2021).

Crude Antigen Preparations for ELISA

Crude antigen preparations are whole schistosome extracts directly extracted from any of the life stages of the parasite. They are the most widely used antigens for ELISA. As diagnostic antigens, these preparations are known to have high sensitivity but low specificity and poor reproducibility (Lv et al., 2019; Zhang et al., 2015).

Soluble Egg Antigen

Among the known antigens used for schistosome antibody detection through ELISA, soluble egg antigen or SEA remains to be the most popular choice. First studied using *Schistosoma mansoni* eggs in 1970 by Boros and Warren, who also coined the term, SEA is a combination of proteins, glycoproteins, and polysaccharides obtained from the clear supernatant of centrifugated homogenized eggs (Boros and Warren, 1970; Chen et al., 2021). It is considered to be one of the main schistosome antigens (Chen et al., 2021; Sarhan et al., 2014). Moreover, SEA – ELISA often serves as the reference immunodiagnostic tool when evaluating other antigens.

SEA is known to give the best sensitivity results among the existing antigens used for ELISA. In a meta-analysis done by Rahman et al., SEA – ELISA showed the highest sensitivity (90-100%) (Rahman et al., 2021). It is also known to have increased sensitivity as the disease progresses (Khalil et al., 1989). In a study by Sarhan et al., on chronic *S. mansoni* infection, SEA – ELISA exhibited an increase in diagnostic efficacy when used in serum of higher concentration (Sarhan et al., 2014). In communities of low *S. mansoni* transmission, *Sm*SEA – ELISA was shown to be the best tool when it comes to diagnosing current and previous infections with a specificity of 65.33% which is further improved to 72% when treated with sodium metaperiodate (SEA-SMP-ELISA) (Ferrer et al., 2020). For documentation of the absence of disease in low transmission areas, SEA – ELISA is recommended (Al – Shehri et al., 2018).

The assay also had better results in diagnosing imported schistosomiasis when compared to AWA – ELISA and CA – ELISA. The in – house assay showed a sensitivity of 90.5% and 57.1% for *S. mansoni* and *S. haematobium* infections, respectively, and specificity of 97.1% for all cases; Moreover, between its commercial counterparts, the ELISA/ DRG performed better which showed a sensitivity of 85.7% and 71.4% for *S. mansoni* and *S. haematobium* infections, respectively, and specificity of 88.4%, as compared to ELISA/ Viramed which showed a sensitivity of 71.4% and 64.3% for *S. mansoni* and *S. haematobium* infections, respectively, and specificity of 88.4% (Kinkel et al., 2012). When compared with urine microscopy (UM), the first-line diagnostic test for urogenital schistosomiasis caused by *S. haematobium*, SEA – ELISA showed a 79.9% positivity rate (sensitivity of 94.8% and specificity of 29.7%) whereas UM had 38.9%with the results for both agreeing 55% of the time making SEA – ELISA a potential supplement tool for UGS diagnosis (Song et al., 2018).

There are some drawbacks, however, in using SEA – ELISA. Making the antigen preparation is laborious as it requires egg collection from laboratory infected host animals (Chand, Chiodini and Doenhoff, 2010). Moreover, it is only recommended to be used for detecting infection up to three months after exposure and exhibits cross-reactivity in cases with *Fasciola and Trichinella* infection (Chand, Chiodini and Doenhoff, 2010; McLaren et al., 1978; Sarhan et al., 2014).

Adult Worm and Cercarial Antigen

One alternative for SEA is the soluble adult worm antigen preparation. AWA represents the complex antigens derived from adult schistosome worms. Compared to eggs, adult schistosome worms offer a higher yield of antigenic products (Hamilton, Klinkert and Doenhoff, 1998).

In a study by Sarhan et al., on diagnosing chronic *S. mansoni* infection, AWA – ELISA showed the highest sensitivity (100%) for both sera dilutions, 1/100 and 1/50, respectively, the lowest specificity (76.47% and 64.7%) for the respective dilutions, and the highest diagnostic efficacy (81.25%) equivalent to that of SEA - ELISA for 1/50 sera dilution (Sarhan et al., 2014).

The assay also performed better than CA – ELISA in diagnosing imported schistosomiasis. The in–house assay showed a sensitivity of 76.2% and 21.4% for *S. mansoni* and *S. haematobium* infections, respectively, and specificity of 100% for all cases, while its commercial counterpart, the ELISA/ NovaTec, showed a sensitivity of 81.0% and 35.7% for *S. mansoni* and *S. haematobium* infections, respectively, and specificity of 99.0% (Kinkel et al., 2012). Moreover, in a study by

Oyeyemi et al., on chronic *S. mansoni* patients from low endemic populations, AWA – ELISA gave the best diagnostic accuracy (93.33) and specificity (96.67) compared to SEA – ELISA and SCA – ELISA (Oyeyemi et al., 2021).

Another alternative is the cercarial antigen (CA). Antibodies to cercaria are known to be elevated during the early phase of schistosome infection making it suitable for detecting early infections (Chand, Chiodini and Doenhoff, 2010; Lunde, Ottesen and Cheever, 1979). However, it has been thought that exposure of the body to CA is relatively short and that the corresponding immune response is downregulated immediately (Mountford and Trottein, 2004). Still, the less time – consuming and less labor – intensive preparation and requirements for obtaining cercarial antigens make CA a reasonable choice.

In a study by Chand et al., among patients in a non – endemic setting, it was shown that the CA – ELISA had similar results with SEA – ELISA in detecting schistosome antibodies with a sensitivity of 99.2% and specificity of 95.3% (Chand, Chiodini and Doenhoff, 2010). Just like SEA – ELISA, CA – ELISA displayed cross-reactivity to *Fasciola* and *Trichinella* infections, and it is only recommended to be used for detecting infection up to three months after exposure (Chand, Chiodini and Doenhoff, 2010; McLaren et al., 1978; Sarhan et al., 2014).

In diagnosing imported schistosomiasis, CA-ELISA performed poorly compared to SEA-ELISA and AWA-ELISA showing a sensitivity of 47.6% and 21.4% for *S. mansoni* and *S. haematobium* infections, respectively, and specificity of 95.2% for all cases (Kinkel et al., 2012).

El – Derbawy et al., evaluated the effect of irradiation on cercarial antigens in diagnosing schistosomiasis. *S. mansoni* cercariae were irradiated with 400 gray gamma (dose rate 2.5 K.Gy/h, intensity 4000 uw/cm^2) for one minute prior to antigen preparation. Irradiated SCA – ELISA showed higher specificity and accuracy (87% and 92.3%) as compared to non – irradiated (83.9% and 90.4%). Both showed a sensitivity of 100%. Irradiated SCA was also noted to have lower cross-reactivity to other parasites (El – Derbawy et al., 2021).

Schistosomula Crude Antigen

Relatively new compared to the crude antigens mentioned above, schistosomula crude antigen (SCA) also has potential. It has been shown to have a significant positivity ratio of 80%, and higher sensitivity than ELISA – SWAP in detecting acute cases of *S. mansoni* infection in travelers (Grenfell et al., 2013). Oyeyemi et al., evaluated the potential use of SCA for detecting *S. mansoni* antibodies through ELISA in chronic patients from low endemic populations. Cercariae were mechanically transformed into schistosomula which were then homogenized and centrifuged to extract the antigens. SCA – ELISA exhibited a higher diagnostic accuracy (91.67%) and sensitivity (91.67%) as compared to SEA – ELISA (88.33% and 90.00%, respectively) (Oyeyemi et al., 2021).

Recombinant Single and Multi - Epitope Antigen Preparations for ELISA

In recent years, with the advent of using more sophisticated equipment and protocols, several potential antigenic targets, especially the tegumental and excretory-secretory proteins, have been identified. This allowed researchers to explore the use of more recombinant antigens in ELISA for the diagnosis of *Schistosoma* infection. The use of purified and single secreted antigens has also helped increase the sensitivity of ELISA, and address the specificity issues of crude antigen preparations (Ongongo, Kariuki and Wilson, 2018; Zhang et al., 2015).

Zhang et al., screened *S. japonicum* antigens among the tegument proteins using an immunoproteomic approach. Among the six proteins identified, SjPGM (phosphoglycerate mutase) and SjRAD23 (UV excision repair protein RAD23 homolog) were expressed by plasmid pET28a (+) in BL21 *E. coli* and were purified for ELISA to detect schistosome antibodies in water buffalos. Results showed that rSjPGM – ELISA and rSjRAD23 – ELISA showed higher specificity (100% and 98.33%, respectively) than SEA – ELISA. Both gave a lower, but acceptable, sensitivity (91.35% and 88.46%, respectively) compared to SEA – ELISA (100.00%) and showed low cross-reactivity to other parasites. Moreover, between the two, rSjPGM – ELISA performed better. They also noted the potential of these antigens in discriminating between active and previous infections since antibodies against them rapidly decreased after praziquantel treatment (Zhang et al., 2015).

In a study by Lv et al., on detecting *S. japonicum* in water buffalos, a trivalent multi-epitope antigen, PA4 (BSjPGM-BSjRAD23-1-BSj23, from SjPGM phosphoglycerate mutase and SjRAD23), recorded the highest sensitivity rate (95.61%) among recombinant antigens with no significant difference compared to SEA (100%, p = 0.836). PA4 also showed better specificity (97.83%) compared to SEA (82.61%) and so did the other recombinant antigens tested. In addition, the recombinant antigens displayed lower cross–reactivity (Lv et al., 2018).

Angeles et al., in a series of studies, assessed the diagnostic potential of thioredoxin peroxidase – 1 (SjTPx-1) and four tandem repeat proteins (Sj1TR, Sj2TR, Sj4TR, Sj7TR) in detecting antibody response against *S. japonicum* in humans, water buffalos, and dogs. For all subjects, SjTPx-1 gave the highest sensitivity (85.71%, 82.61%, and 80%, respectively) among the recombinant candidates, making it a potentially good antigen. In addition, it showed very minimal cross-reactivity to other helminths. Sj7TR was good for humans and dogs and showed a sensitivity of 73.3% and 85.71%, and a specificity of 92.3% and 100%, respectively, while Sj1TR, on the other hand, was good for water buffalos and showed a sensitivity of 78.26% and specificity of 93.33% (Angeles et al., 2011; Angeles et al., 2012; Angeles et al., 2019). The results for Sj1TR were further supported by field trials among water buffalos in the Philippines since it gave a higher specificity, sensitivity, and accuracy (83.33%, 97.35%, and 92.81%, respectively), as compared to rSjTPx-1 (68.52%, 87.61 and 81.44%, respectively), and it even outperformed SEA – ELISA (70.37%, 69.03%, and 69.46%, respectively) (Angeles et al., 2020).

Cai et al., compared ten recombinant antigen candidates when used in ELISA for diagnosing *S. japonicum* Philippine strain (SjP) in 180 patients. DNA fragments from a region of the target proteins were amplified and cloned into the pET – 28a vector. Results showed SjSAP4, from a secretory protein, had the highest diagnostic efficacy with a sensitivity of 94.03% and specificity of 98.3% for patients with high infection intensity. Moreover, SjSAP4 + Sj23 – LHD, SjSAP4 with an attached large hydrophilic domain, showed the highest diagnostic efficacy with a sensitivity of 87.04% and specificity of 96.67% for patients with low infection intensity (Cai et al., 2017). These results were supported by another study by Cai et al., in a setting with moderate prevalence but low-intensity infection where SjSAP4 + Sj23 – LHD showed the highest specificity (92.1%) and diagnostic accuracy (67.2%) among the ELISA antigens (Cai et al, 2019).

Another potential diagnostic antigen is the *S. japonicum* cathepsin B (SjCatB), a protease used by the cercaria for penetrating the host's skin. Macalanda et al., cloned and expressed recombinant SjCatB and evaluated its diagnostic potential thru ELISA with sera from laboratory infected mice. Results showed a sensitivity and specificity of 100% and 95.0%, respectively, and that antibodies could be detected as early as six weeks post-infection (in 2 out of 3 infected mice) suggesting its potential use for diagnosing early infections. When tested with sera from individuals infected with *S. japonicum,* rSjCatB – ELISA gave a sensitivity of 86.7% and specificity of 96.7% and a very minimal cross-reactivity with other parasitic infections (Macalanda et al., 2019).

Dang – Trinh et al., evaluated the potential of peroxiredoxins, particularly *Sj*Prx – 4, in diagnosing *S. japonicum* infection. The antigen was subcloned from PCR 2.1 – TOPO vector into pGEX – 6P – 1 vector, expressed through *E. coli* strain Rosetta (DE3), and used in ELISA (r*Sj*Prx – 4 – ELISA) for detection of schistosome antibodies in infected humans. Recombinant *Sj*Prx – 4 – ELISA gave a higher specificity (87.7%) but lower sensitivity (83.3) compared to SEA – ELISA (70.7% and 90%, respectively). When combined with *Sj*TPx – 1 (*Sj*Prx – 4/ *Sj*TPx – 1), its sensitivity and specificity were improved to 90% and 92.7%, respectively, equal and higher than that of SEA (Dang – Trinh et al., 2020). These results showed the potential of using fusion antigens in diagnostics.

A study by Tanaka et al., further explored the potential of SmSerpin (*S. mansoni* serine protease – inhibitor) and RP26 for detection of *S. mansoni* antibody in patients of light infection intensity through ELISA. Both antigens and their cocktail showed higher specificity (62.4%, 57.4%, and 61.4%, respectively) but lower sensitivity (65.3%, 74.5%, and 83.7%, respectively) than SEA (29.7% and 90.8%, respectively). The improved sensitivity of the cocktail antigen made it a potential alternative for diagnosing schistosomiasis in low-endemic areas (Tanaka et al., 2021).

Some recently identified antigens are still being explored and on their preliminary evaluation. In 2017, SjMRP1, multidrug resistance protein 1 gene, was identified by Feng et al., through screening a cDNA phage display library from *S. japonicum* adult worms and was cloned and expressed as a recombinant protein, rSjMRP1, for evaluation thru ELISA (rSjMRP1 – ELISA) in domestic animals, particularly goat, cattle and buffalo. Results were not significantly different from that of SEA – ELISA and showed a sensitivity of 95.6%, 100%, and 90%, for the respective animals, and specificity of 100%, 100%, and 90%, respectively. Cross-reactivity was not observed with other parasites in the study (Feng et al., 2017).

Silva – Moraes et al., through immunoproteomic analyses of egg extracts, identified a new antigen candidate, Major Egg Antigen (MEA). Using Gateway cloning technology (Invitrogen), the protein was cloned, expressed, and purified to produce the recombinant antigen, rMEA. The antigen was then used in IgG ELISA to diagnose *S. mansoni* infections of low intensity in an endemic setting. rMEA – ELISA gave a sensitivity of 87.10% and specificity of 89.09%. In addition, rMEA – ELISA was able to detect infections missed by Kato – Katz (Silva – Moraes et al., 2019).

Chen et al., assessed the diagnostic potential of recombinant schistosome-specific extracellular vesicle proteins, and identified two peptides, from an uncharacterized protein and a cathepsin B – like cysteine proteinase, which had detectable antibody levels in *S. japonicum* infected humans (Chen et al., 2020).

Magnetic Affinity Enzyme-Linked Immunoassay

The search for a more sensitive screening method for schistosomiasis has driven researchers to find ways to innovate ELISA. Through the years, several variations and adjunct features have been added to the assay to address its limitations.

Paramagnetic beads are magnetic particles coated with ligands that specifically bind to a target molecule. These beads have previously been used in the diagnosis of several parasitic diseases particularly in schistosomiasis through magnetic affinity enzyme-linked immunoassay (MEIA) (Chen et al., 2020; Teixeira et al., 2007). When used in immunoassays, they are able to improve sensitivity and make antibody binding more efficient (Hayat, Barthelmebs and Marty, 2011; Yu et al., 2012a).

In 2012, Yu et al., developed a MEIA for diagnosing schistosomiasis in low infection intensity patients based on a number of antigens including recombinant 26 kDa glutathione – S – transferase (rSj26GST), and signal transduction protein 14 – 3 – 3 (rSj14 – 3 – 3). Antigens are cloned in pET28a and expressed in *E. coli* BL21 (DE3). For rSj26GST, results showed that rSj26GST – MEIA and rSj26GST – ELISA gave similar positive detection rates (5/28). When it came to the ratio of mean positive value to the mean negative value (P/N), MEIA gave a higher score than ELISA (3.13 and 2.24, respectively). When used to evaluate treated patients, MEIA and ELISA gave similar sensitivity (2/15) and negative rate (86.67%) (Yu et al., 2012a). On the other hand, rSj14 – 3 – 3 – MEIA gave a higher positive detection rate (13/58) compared to rSj14 – 3 – 3 – ELISA (10/58). It also showed higher P/N (3.57) than ELISA (2.68). Both recombinant antigens, whether MEIA or ELISA, had no cross-reactivity to paragonimiasis (Yu et al., 2012a; Yu et al., 2013).

To evaluate MEIA further, Yu et al., developed a MEIA based on crude antigen SEA for diagnosing schistosomiasis in patients with low-intensity infection.

SEA – MEIA gave a statistically significant higher positivity rate of 96.55% and P/N of 4.42 compared to SEA – ELISA with 91.38% and 3.73, respectively. MEIA also performed better in giving negative results for patients treated for six months. However, unlike the recombinant antigens tested, SEA, whether MEIA or ELISA, showed cross-reactivity to paragonimiasis (50%) (Yu et al., 2012b).

Conclusion and Future Perspectives

The use of ELISA for the diagnosis of schistosomiasis has indeed improved throughout the years. Table 1 summarizes the diagnostic capabilities of various types of ELISA using either crude or recombinant antigens for the diagnosis of schistosomiasis. Even in the presence of its known limitations or the rise of nucleic acid-based techniques, ELISA remains a necessary tool needed for surveillance and control of the disease.

Although finding new antigens is valuable, attention to enhancing the sensitivity and, especially, specificity of ELISA is as important. This will lead to early detection of infection and, consequently, early therapeutic intervention. Moreover, finding and developing antigen candidates that can distinguish current and previous infections is also important since they can help with evaluating treatment success. Large-scale testing is also needed for a better assessment of the quality of newly evaluated antigens.

Diagnosis plays a central role in the control and management of schistosomiasis. That said, there is always a need to strengthen and improve schistosome diagnostics. Combining the coproparasitology and serology–based techniques in diagnostics may address the limitations of the two and make surveillance and control more efficient. With the right diagnostic tools and global cooperation, only then will the common goal of successful elimination of schistosomiasis be actualized.

Table 1. Summary of antigens used for antibody detection in diagnosing schistosomiasis

Antigenic Origin	Antigen	Technique	Sensitivity (%)	Specificity (%)	Remarks[a]	Ref.[b.]
Egg	*Sj*SEA	ELISA	97.14	71.76	*S. japonicum* infection in an endemic setting	Angeles et al., 2011
	*Sj*SEA	ELISA	69.57	86.67	Water buffalo *S japonicum* infection in an endemic setting	Angeles et al., 2012
	*Sj*SEA	ELISA	80	59.6	Canine *S japonicum* infection in an endemic setting	Angeles et al., 2019
	*Sj*SEA	ELISA	70.37	69.03	Field evaluation for water buffalo *S japonicum* infection an in endemic setting	Angeles et al., 2020
	*Sj*SEA	ELISA	90	70.7	*S. japonicum* infection in an endemic setting	Dang – Trinh et al., 2020
	*Sj*SEA	ELISA	100 95.45 100	84.33 100 100	For goat, cattle, and buffalo, respectively; *S. japonicum* laboratory infection	Feng et al., 2017
	*Sm*SEA	ELISA	47.22 38.89	65.33 72.00	For normal and SMP – coupled SEA, respectively; *S. mansoni* infection in low transmission community	Ferrer et al., 2020
	*Sm*SEA	ELISA	90.5 57.1	97.1 (for all cases)	Imported *S. mansoni* and *S. haematobium* infection, respectively	Kinkel et al., 2012
	*Sj*SEA	ELISA	100	82.6	Water buffalo *S japonicum* infection in an endemic setting	Lv et al., 2018
	*Sm*SEA	ELISA	90	86.67	Chronic *S. mansoni* in a low endemic setting	Oyeyemi et al., 2021
	*Sj*SEA	ELISA	90 – 100	23 – 61	Meta-analysis; Endemic countries of *S. japonicum*	Rahman et al., 2021
	*Sm*SEA	ELISA	80 80	76.47 82.35	For 1/100 and 1/50 sera dilutions, respectively; chronic *S. mansoni* infection in an endemic setting	Sarhan et al., 2014

Table 1. (Continued)

Antigenic Origin	Antigen	Technique	Sensitivity (%)	Specificity (%)	Remarks[a]	Ref.[b.]
	*Sh*SEA	ELISA	98.4	29.7	*S. haematobium* infection in an endemic setting	Song et al., 2018
	*Sm*SEA	ELISA	90.8	29.7	*S. mansoni* infection in low infection intensity in a low endemic setting	Tanaka et al., 2021
	*Sj*SEA	MEIA ELISA	96.5 91.1	100 100	Low-intensity *S. japonicum* infection	Yu et al., 2012b
	*Sj*SEA	ELISA	100	91.67	Water buffalo *S japonicum* infection in an endemic setting	Zhang et al., 2015
Cercaria	*Sm*CA	ELISA	99.2	95.3	*S. mansoni* and *S. haematobium* infection in non – endemic setting	Chand, Chiodini and Doenhoff, 2010
	*Sm*CA	ELISA	100 100	87 83.9	For irradiated and normal CA, respectively; *S. mansoni* infection in an endemic setting	El – Derbawy et al., 2021
	*Sm*CA	ELISA	47.6 21.4	95.2 (for all cases)	Imported *S. mansoni* and *S. haematobium* infection, respectively	Kinkel et al., 2012
	*Sm*CA	ELISA	86.6 80	94.11 76.47	For 1/100 and 1/50 sera dilutions, respectively; chronic *S. mansoni* infection in an endemic setting	Sarhan et al., 2014
Schisoso-mulum	*Sm*SCA	ELISA	96.67	86.67	Chronic *S. mansoni* in a low endemic setting	Oyeyemi et al., 2021
Adult Worm	*Sm*AWA	ELISA	76.2 21.4	100 (for all cases)	Imported *S. mansoni* and *S. haematobium* infection, respectively	Kinkel et al., 2012
	*Sm*AWA	ELISA	90	96.67	Chronic *S. mansoni* in a low endemic setting	Oyeyemi et al., 2021
	*Sm*AWA	ELISA	100 100	76.4 64.7	For 1/100 and 1/50 sera dilutions, respectively; chronic *S. mansoni* infection in an endemic setting	Sarhan et al., 2014

Antigenic Origin	Antigen	Technique	Sensitivity (%)	Specificity (%)	Remarks[a]	Ref.[b.]
Recombinant	*Sj*TPx – 1	ELISA	85.71	96.47	*S. japonicum* infection in an endemic setting	Angeles et al., 2011
	*Sj*1TR		68.57	100		
	*Sj*4TR		57.14	98.82		
	*Sj*7TR		85.71	100		
	*Sj*TPx – 1	ELISA	82.61	96.67	Water buffalo *S japonicum* infection an in endemic setting	Angeles et al., 2012
	*Sj*1TR		78.26	93.33		
	*Sj*2TR		56.52	96.67		
	*Sj*4TR		60.87	100		
	*Sj*7TR		60.87	100		
	*Sj*TPx – 1	ELISA	80	92.3	Canine *S japonicum* infection on in endemic setting	Angeles et al., 2019
	*Sj*1TR		53.3	88.5		
	*Sj*2TR		40	94.2		
	*Sj*4TR		60	94.2		
	*Sj*7TR		73.3	92.3		
	*Sj*TPx – 1	ELISA	68.52	87.61	Field evaluation for water buffalo *S japonicum* infection in an endemic setting	Angeles et al., 2020
	*Sj*1TR		83.33	97.35		
	*Sj*SAP4	ELISA	94.03	98.33	Small scale cohort of *S. japonicum* infection in an endemic setting	Cai et al., 2017
	*Sj*SAP4 + *Sj*23 – LHD		91.04	98.33		
	*Sj*SAP4 + *Sj*23 – LHD	ELISA	87.04	96.67	Large scale cohort of *S. japonicum* infection in an endemic setting	
	*Sj*SAP4		84.26	96.67		
	*Sj*23 – LHD	ELISA	50.9	56.3	*S. japonicum* infection in a low infection intensity setting	Cai et al., 2019
	*Sj*SAP4		78.3	33.3		
	*Sj*SAP5		79.4	18.5		
	*Sj*SAP4 + *Sj*23 – LHD		76.9	36.3		
	*Sj*SAP5 + *Sj*23 – LHD		65.7	37.8		
	*Sj*Prx – 4	ELISA	83.3	97.6	*S. japonicum* infection in an endemic setting	Dang – Trinh et al., 2020
	*Sj*TPx – 1		83.3	97.6		
	*Sj*TPx – 1/ *Sj*Prx – 4		90	92.7		

Table 1. (Continued)

Antigenic Origin	Antigen	Technique	Sensitivity (%)	Specificity (%)	Remarks[a]	Ref.[b.]
	*Sj*MRP – 1	ELISA	95.6 100 90	100 100 96.67	For goat, cattle, and buffalo, respectively; *S. japonicum* laboratory infection	Feng et al., 2017
	PA1	ELISA	48.25	97.83	Monovalent BSjRAD23 – 1; Water buffalo *S japonicum* infection in an endemic setting	Lv et al., 2018
	PA2	ELISA	82.46	92.39	Monovalent BSj23; Water buffalo *S japonicum* infection in an endemic setting	
	PA3	ELISA	70.18	98.91	Bivalent BSjPGM – BSj23; Water buffalo *S japonicum* infection in an endemic setting	
	PA4	ELISA	95.61	97.83	Trivalent; Water buffalo *S japonicum* infection in an endemic setting	
	PA5	ELISA	67.54	100	Trivalent BSjRAD23 – 2 – BSjPGM – BSj23; Water buffalo *S japonicum* infection in an endemic setting	
	*Sj*CatB	ELISA	86.7	96.7	*S. japonicum* infection in an endemic setting	Macalanda et al., 2019
	*Sm*MEA	ELISA	87.10	89.09	*S. mansoni* infection in low infection intensity in a low endemic setting	Silva – Moraes et al., 2019
	*Sm*Serpin	ELISA	65.3	62.4	*S. mansoni* infection in low infection intensity in a low endemic setting	Tanaka et al., 2021
	*Sm*RP26		74.5	57.4		
	*Sm*Serpin/ *Sm*RP26		83.7	61.4		
	*Sj*14 – 3 – 3	MEIA ELISA	22.4 17.2	100 100	Low-intensity *S. japonicum* infection	Yu et al., 2012a
	*Sj*26GST	MEIA ELISA	17.8 17.8	100 100		
	SjPGM	*ELISA*	91.35	100		
	SjRAD23	*ELISA*	88.46	98.33		

[a]Serum samples are from human subjects unless specified.

[b]Entries per antigenic origin are arranged in alphabetical order according to references.

REFERENCES

Angeles, J. M., Goto, Y., Kirinoki, M., Asada, M., Leonardo, L. R., Rivera, P. T., Villacorte, E. A., Inoue, N., Chigusa, Y., and Kawazu, S. (2012). Utilization of ELISA using thioredoxin peroxidase-1 and tandem repeat proteins for diagnosis of *Schistosoma japonicum* infection among water buffaloes. *PLoS Neglected Tropical Diseases*, 6 (8), e1800. https://doi.org/10.1371/journal.pntd.0001800.

Angeles, J., Goto, Y., Kirinoki, M., Leonardo, L. R., Moendeg, K. J., Ybañez, A. P., Rivera, P. T., Villacorte, E. A., Inoue, N., Chigusa, Y., and Kawazu, S. I. (2019). Detection of canine *Schistosoma japonicum* infection using recombinant thioredoxin peroxidase-1 and tandem repeat proteins. *The Journal of Veterinary Medical Science*, 81(10); 1413–1418. https://doi.org/10.1292/jvms.19-0126.

Angeles, J. M., Goto, Y., Kirinoki, M., Leonardo, L., Tongol-Rivera, P., Villacorte, E., Inoue, N., Chigusa, Y., and Kawazu, S. (2011). Human antibody response to thioredoxin peroxidase-1 and tandem repeat proteins as immunodiagnostic antigen candidates for *Schistosoma japonicum* infection. *The American Journal of Tropical Medicine and Hygiene*, 85(4): 674–679. https://doi.org/10.4269/ajtmh.2011.11-0245.

Angeles, J., Goto, Y., Kirinoki, M., Villacorte, E. A., Moendeg, K. J., Rivera, P. T., Chigusa, Y., and Kawazu, S. I. (2020). Field evaluation of recombinant antigen ELISA in detecting zoonotic schistosome infection among water buffaloes in endemic municipalities in the Philippines. *Frontiers on Veterinary Science*, 7: 592783. https://doi.org/10.3389/fvets.2020.592783.

Ajibola, O., Gulumbe, B. H., Eze, A. A., and Obishakin, E. (2018). Tools for detection of schistosomiasis in Resource Limited Settings. *Medical Sciences* (Basel, Switzerland), 6(2): 39. https://doi.org/10.3390/medsci6020039.

Al-Shehri, H., Koukounari, A., Stanton, M. C., Adriko, M., Arinaitwe, M., Atuhaire, A., Kabatereine, N. B., and Stothard, J. R. (2018). Surveillance of intestinal schistosomiasis during control: a comparison of four diagnostic tests across five Ugandan primary schools in the Lake Albert region. *Parasitology*, 145(13): 1715–1722. https://doi.org/10.1017/S003118201800029X.

Bergquist, R., and Gray, D. J. (2019). Schistosomiasis elimination: Beginning of the end or a continued march on a trodden path. *Tropical medicine and infectious disease*, 4(2): 76. https://doi.org/10.3390/tropicalmed4020076.

Blas, B. L., Rosales, M. I., Lipayon, I. L., Yasuraoka, K., Matsuda, H., and Hayashi, M. (2004). The schistosomiasis problem in the Philippines: a review. *Parasitology International*, 53(2): 127–134. https://doi.org/10.1016/j.parint.2004.01.003.

Boros, D. L., and Warren, K. S. (1970). Delayed hypersensitivity-type granuloma formation and dermal reaction induced and elicited by a soluble factor isolated from *Schistosoma mansoni* eggs. *The Journal of Experimental Medicine*, 132(3): 488–507. https://doi.org/10.1084/jem.132.3.488.

Cai, P., Weerakoon, K. G., Mu, Y., Olveda, D. U., Piao, X., Liu, S., Olveda, R. M., Chen, Q., Ross, A. G., and McManus, D. P. (2017). A parallel comparison of antigen candidates for development of an optimized serological diagnosis of schistosomiasis japonica in the Philippines. *EBioMedicine*, 24: 237–246. https://doi.org/10.1016/j.ebiom.2017.09.011.

Cai, P., Weerakoon, K. G., Mu, Y., Olveda, R. M., Ross, A. G., Olveda, D. U., and McManus, D. P. (2019). Comparison of Kato Katz, antibody-based ELISA and droplet digital PCR diagnosis of schistosomiasis japonica: Lessons learnt from a setting of low infection intensity. *PLoS Neglected Tropical Diseases*, 13(3): e0007228. https://doi.org/10.1371/journal.pntd.0007228.

Cesari, I. M., Ballen, D. E., Mendoza, L., and Matos, C. (2005). Detection of *Schistosoma mansoni* membrane antigens by immunoblot analysis of sera of patients from low-transmission areas. *Clinical and Diagnostic Laboratory Immunology*, 12(2): 280–286. https://doi.org/10.1128/CDLI.12.2.280-286.2005.

Chand, M. A., Chiodini, P. L., and Doenhoff, M. J. (2010). Development of a new assay for the diagnosis of schistosomiasis, using cercarial antigens. *Transactions of the Royal Society of Tropical Medicine and Hygiene*, 104(4): 255–258. https://doi.org/10.1016/j.trstmh.2009.12.004.

Chen, Y., Giri, B. R., Li, X., He, X., Jing, Z., and Cheng, G. (2020). Preliminary evaluation of the diagnostic potential of *Schistosoma japonicum* extracellular vesicle proteins for *Schistosomiasis japonica. Acta Tropica*, 201: 105184. https://doi.org/10.1016 /j.actatropica. 2019.105184.

Chen, C., Guo, Q., Fu, Z., Liu, J., Lin, J., Xiao, K., Sun, P., Cong, X., Liu, R., and Hong, Y. (2021). Reviews and advances in diagnostic research on *Schistosoma japonicum. Acta Tropica*, 213: 105743. https://doi.org/10.1016/j.actatropica.2020.105743.

Dang-Trinh, M. A., Angeles, J., Moendeg, K. J., Macalanda, A., Nguyen, T. T., Higuchi, L., Nakagun, S., Kirinoki, M., Chigusa, Y., Goto, Y., and Kawazu, S. I. (2020). Analyses of the expression, immunohistochemical properties and serodiagnostic potential of *Schistosoma japonicum* peroxiredoxin-4. *Parasites & Vectors*, 13(1): 436. https://doi.org/10.1186/s13071-020-04313-w.

Doenhoff, M. J., Chiodini, P. L., and Hamilton, J. V. (2004). Specific and sensitive diagnosis of schistosome infection: can it be done with antibodies?. *Trends in Parasitology*, 20(1): 35–39. https://doi.org/10.1016/j.pt.2003.10.019.

El – Derbawy, M. M., Hafez, E. N., Rabbo, M. A. D. A., Toaleb, N. I., El – Ahl, S. A. H., El – Aswad, B. E. W. (2021). Assessment of the diagnostic performance of normal and irradiated *Schistosoma mansoni* cercarial antigens using ELISA. *Pakistan Journal on Zoology*, 1-6.

El – kady, A. M., Abdella, O. H., Hassan, A. A., Mokhtar, Y., and Hassan, M. H. (2018). Comparative validity of ELISA and indirect haemagglutination in diagnosing *Schistosoma haematobium* infection: an Egyptian study. *International Journal of Tropical Medicine*, 13: 4 – 6.

Espírito-Santo, M. C., Alvarado-Mora, M. V., Dias-Neto, E., Botelho-Lima, L. S., Moreira, J. P., Amorim, M., Pinto, P. L., Heath, A. R., Castilho, V. L., Gonçalves, E. M., Luna, E. J., Carrilho, F. J., Pinho, J. R., and Gryschek, R. C. (2014). Evaluation of real-time PCR assay to detect *Schistosoma mansoni* infections in a low endemic setting. *BMC Infectious Diseases*, 14: 558. https://doi.org/10.1186/s12879-014-0558-4.

Feng, J., Xu, R., Zhang, X., Han, Y., He, C., Lu, C., Hong, Y., Lu, K., Li, H., Jin, Y., Lin, J., and Liu, J. (2017). A candidate recombinant antigen for diagnosis of *schistosomiasis japonica* in domestic animals. *Veterinary Parasitology*, 243: 242–247. https://doi.org/10.1016/j.vetpar.2017.06.015.

Ferrer, E., Villegas, B., Mughini-Gras, L., Hernández, D., Jiménez, V., Catalano, E., and Incani, R. N. (2020). Diagnostic performance of parasitological, immunological and molecular tests for the diagnosis of *Schistosoma mansoni* infection in a community of low transmission in Venezuela. *Acta Tropica*, 204: 105360. https://doi.org/10.1016/j.actatropica.2020.105360.

Francisco, I., Jiz, M., Rosenbaum, M., Baltazar, P., and Steele, J. A. (2019). Knowledge, attitudes and practices related to schistosomiasis transmission and control in Leyte, Philippines. *PLoS Neglected Tropical Diseases*, 13(5): e0007358. https://doi.org/10.1371/journal.pntd.0007358.

Gordon, C. A., Kurscheid, J., Williams, G. M., Clements, A., Li, Y., Zhou, X. N., Utzinger, J., McManus, D. P., and Gray, D. J. (2019). Asian schistosomiasis: current status and prospects for control leading to elimination. *Tropical Medicine and Infectious Disease*, 4(1): 40. https://doi.org/10.3390/tropicalmed4010040.

Grenfell, R. F., Martins, W., Drummond, S. C., Antunes, C. M., Voieta, I., Otoni, A., Oliveira, A. A., Silva-Moraes, V., Oliveira, E. R., Oliveira, E., Lambertucci, J. R., Fonseca, C. T., and Coelho, P. M. (2013). Acute schistosomiasis diagnosis: a new tool for the diagnosis of schistosomiasis in a group of travelers recently infected in a new focus of *Schistosoma mansoni*. *Revista da Sociedade Brasileira de*

Medicina Tropical, 46(2): 208–213. https://doi.org/10.1590/0037-8682-0064-2012.

Hamilton, J. V., Klinkert, M., and Doenhoff, M. J. (1998). Diagnosis of schistosomiasis: antibody detection, with notes on parasitological and antigen detection methods. *Parasitology*, 117 Suppl: S41–S57. https://doi.org/10.1017/s0031182099004205.

Hayat, A., Barthelmebs, L., and Marty, J. L. (2011). Enzyme-linked immunosensor based on super paramagnetic nanobeads for easy and rapid detection of okadaic acid. *Analytica Chimica Acta*, 690, 248 – 252.

Huldt, G., Lagerquist, B., Phillips, T., Draper, C. C., and Voller, A. (1975). Detection of antibodies in schistosomiasis by enzyme-linked immunosorbent assay (ELISA). *Annals of Tropical Medicine & Parasitology*, 69 (4): 483 – 488.

Khalil, H. M., Makled, M. K., Elmissiry, A. G., Khalil, N. M., and Sonbol, S. E. (1989). The application of *Schistosoma mansoni* adult and soluble egg antigens for the serodiagnosis of schistosomiasis by CIEP, IHA and ELISA. *Journal of the Egyptian Society of Parasitology*, 19(2 Suppl): 827–843.

Kinkel, H. F., Dittrich, S., Bäumer, B., and Weitzel, T. (2012). Evaluation of eight serological tests for diagnosis of imported schistosomiasis. *Clinical and Vaccine Immunology*, 19(6): 948–953. https://doi.org/10.1128/CVI.05680-11.

Liu, Z., Zhang, L., Yang, H., Zhu, Y., Jin, W., Song, Q., and Yang, X. (2010). Magnetic microbead-based enzyme-linked immunoassay for detection of *Schistosoma japonicum* antibody in human serum. *Analytical Biochemistry*, 404(2): 127–134. https://doi.org/10.1016/j.ab.2010.05.006.

Lunde, M. N., Ottesen, E. A., and Cheever, A. W. (1979). Serological differences between acute and chronic schistosomiasis mansoni detected by enzyme-linked immunosorbent assay (ELISA). *The American Journal of Tropical Medicine and Hygiene*, 28(1): 87–91. https://doi.org/10.4269/ajtmh.1979.28.87.

Lv, C., Fu, Z., Lu, K., Yue, R., Wang, T., Cao, X., Zhu, C., Li, H., Hong, Y., and Lin, J. (2018). A perspective for improving the sensitivity of detection: The application of multi-epitope recombinant antigen in serological analysis of buffalo schistosomiasis. *Acta Tropica*, 183: 14–18. https://doi.org/10.1016/j.actatropica.2018.03.025.

Macalanda, A., Angeles, J., Moendeg, K. J., Dang-Trinh, M. A., Higuchi, L., Kirinoki, M., Chigusa, Y., Leonardo, L. R., Villacorte, E. A., Rivera, P. T., Goto, Y., and Kawazu, S. I. (2019). *Schistosoma japonicum* cathepsin B as potential diagnostic antigen for Asian zoonotic schistosomiasis. *Parasitology Research*, 118(9): 2601–2608. https://doi.org/10.1007/s00436-019-06410-x.

McLaren, M., Draper, C. C., Roberts, J. M., Minter-Goedbloed, E., Ligthart, G. S., Teesdale, C. H., Amin, M. A., Omer, A. H., Bartlett, A., and Voller, A. (1978). Studies on the enzyme linked immunosorbent assay (ELISA) test for *Schistosoma mansoni* infections. *Annals of Tropical Medicine and Parasitology*, 72(3): 243–253. https://doi.org/10.1080/00034983.1978.11719312.

McManus, D. P., Dunne, D. W., Sacko, M., Utzinger, J., Vennervald, B. J., and Zhou, X. N. (2018). Schistosomiasis. *Nature Reviews. Disease Primers*, 4(1): 13. https://doi.org/10.1038/s41572-018-0013-8.

Mountford, A. P., and Trottein, F. (2004). Schistosomes in the skin: a balance between immune priming and regulation. *Trends in Parasitology*, 20(5): 221–226. https://doi.org/10.1016/j.pt.2004.03.003.

Olveda, R. M., and Gray, D. J. (2019). Schistosomiasis in the Philippines: Innovative control approach is needed if elimination is the goal. *Tropical Medicine and Infectious Disease*, 4(2): 66. https://doi.org/10.3390/tropicalmed4020066.

Ogongo, P., Kariuki, T. M., and Wilson, R. A. (2018). Diagnosis of schistosomiasis mansoni: an evaluation of existing methods and research towards single worm pair detection. *Parasitology*, 145(11): 1355–1366. https://doi.org/10.1017/S0031182018000240.

Oyeyemi, O. T., Corsini, C. A., Goncalves, G., Borges, W. dC., and Grenfell, R. F. Q. (2021). Evaluation of schistosomula crude antigen

(SCA) as a diagnostic tool for *Schistosoma mansoni* in low endemic human population. *Scientific Reports*, 11(1): 10530. https://doi.org/10.1038/s41598-021-89929-3.

Rahman, M. O., Sassa, M., Parvin, N., Islam, M. R., Yajima, A., and Ota, E. (2021). Diagnostic test accuracy for detecting *Schistosoma japonicum* and *S. mekongi* in humans: A systematic review and meta-analysis. *PLoS Neglected Tropical Diseases*, 15(3): e0009244. https://doi.org/10.1371/journal.pntd.0009244.

Rudge, J. W., Carabin, H., Balolong, E., Tallo, V., Shrivastava, J., Lu, D. B., Basáñez, M. G., Olveda, R., McGarvey, S. T., and Webster, J. P. (2008). Population genetics of *Schistosoma japonicum* within the Philippines suggest high levels of transmission between humans and dogs. *PLoS Neglected Tropical Diseases*, 2(11): e340. https://doi.org/10.1371/journal.pntd.0000340.

Sarhan, R. M., Aminou, H. A., Saad, G. A., and Ahmed, O. A. (2014). Comparative analysis of the diagnostic performance of adult, cercarial and egg antigens assessed by ELISA, in the diagnosis of chronic human *Schistosoma mansoni* infection. *Parasitology Research*, 113(9): 3467–3476. https://doi.org/10.1007/s00436-014-4017-3.

Silva-Moraes, V., Shollenberger, L. M., Castro-Borges, W., Rabello, A., Harn, D. A., Medeiros, L., Jeremias, W. J., Siqueira, L., Pereira, C., Pedrosa, M., Almeida, N., Almeida, A., Lambertucci, J. R., Carneiro, N., Coelho, P., and Grenfell, R. (2019). Serological proteomic screening and evaluation of a recombinant egg antigen for the diagnosis of low-intensity *Schistosoma mansoni* infections in endemic area in Brazil. *PLoS Neglected Tropical Diseases*, 13(3): e0006974. https://doi.org/10.1371/journal.pntd.0006974.

Song, H. B., Kim, J., Jin, Y., Lee, J. S., Jeoung, H. G., Lee, Y. H., Saeed, A., and Hong, S. T. (2018). Comparison of ELISA and urine microscopy for diagnosis of *Schistosoma haematobium. Infection. Journal of Korean Medical Science*, 33(33): e238. https://doi.org/10.3346/jkms.2018.33.e238.

Tanaka, M., Kildemoes, A. O., Chadeka, E. A., Cheruiyot, B. N., Sassa, M., Moriyasu, T., Nakamura, R., Kikuchi, M., Fujii, Y., de Dood, C. J., Corstjens, P., Kaneko, S., Maruyama, H., Njenga, S. M., de Vrueh, R., Hokke, C. H., and Hamano, S. (2021). Potential of antibody test using *Schistosoma mansoni* recombinant serpin and RP26 to detect light-intensity infections in endemic areas. *Parasitology International*, 83: 102346. https://doi.org/10.1016/j.parint.2021.102346.

Fagundes Teixeira, C., Neuhauss, E., Ben, R., Romanzini, J., and Graeff-Teixeira, C. (2007). Detection of *Schistosoma mansoni* eggs in feces through their interaction with paramagnetic beads in a magnetic field. *PLoS Neglected Tropical Diseases*, 1(2): e73. https://doi.org/10.1371/journal.pntd.0000073.

Tianping, W., and Jiaojiao, L. (2005). "Animal reservoirs for Schistosoma japonicum in China." In *Asian Parasitology: Schistosomiasis in Asia (Vol. 5)* edited by M. Chen, X. Zhou, K. Hirayama, 151 – 162. Chiba, Japan: AAA Committee.

Turner, P., Lalloo, K., Bligh, J., Armstrong, M., Whitty, C. J., Doenhoff, M. J., and Chiodini, P. L. (2004). Serological speciation of human schistosome infections by ELISA with a panel of three antigens. *Journal of Clinical Pathology*, 57(11): 1193–1196. https://doi.org/10.1136/jcp.2003.014779.

Utzinger, J., Becker, S. L., van Lieshout, L., van Dam, G. J., and Knopp, S. (2015). New diagnostic tools in schistosomiasis. *Clinical Microbiology and Infection: the official publication of the European Society of Clinical Microbiology and Infectious Diseases*, 21(6): 529–542. https://doi.org/10.1016/j.cmi.2015.03.014.

Wang, W., Li, Y., Li, H., Xing, Y., Qu, G., Dai, J., and Liang, Y. (2012). Immunodiagnostic efficacy of detection of *Schistosoma japonicum* human infections in China: a meta analysis. *Asian Pacific Journal of Tropical Medicine*, 5(1): 15–23. https://doi.org/10.1016/S1995-7645(11)60238-1.

Weerakoon, K. G., Gobert, G. N., Cai, P., and McManus, D. P. (2015). Advances in the diagnosis of human schistosomiasis. *Clinical*

Microbiology Reviews, 28(4): 939–967. https://doi.org/10.1128/CMR.00137-14.

Weerakoon, K. G., Gordon, C. A., and McManus, D. P. (2018). DNA diagnostics for schistosomiasis control. *Tropical Medicine and Infectious Disease*, 3(3): 81. https://doi.org/10.3390/tropicalmed3030081.

World Health Organization. (2018). Schistosomiasis and soil-transmitted helminthiases: numbers of people treated in 2017. *Weekly Epidemiological Record*, 93, 681 – 692.

Yu, Q., Yang, H., Feng, Y., Yang, X., and Zhu, Y. (2012). Magnetic affinity enzyme-linked immunoassay based on recombinant 26 kDa glutathione-S-transferase for serological diagnosis of schistosomiasis japonica. *Acta Tropica*, 124(3): 199–202. https://doi.org/10.1016/j.actatropica.2012.08.006.

Yu, Q., Yang, H., Feng, Y., Zhu, Y., and Yang, X. (2012). Magnetic affinity enzyme-linked immunoassay for diagnosis of schistosomiasis japonicum in persons with low-intensity infection. *The American Journal of Tropical Medicine and Hygiene*, 87(4): 689–693. https://doi.org/10.4269/ajtmh.2012.11-0716.

Yu, Q., Yang, H., Guan, F., Feng, Y., Yang, X., and Zhu, Y. (2014). Detection of IgG in sera of patients with schistosomiasis japonica by developing magnetic affinity enzyme-linked immunoassay based on recombinant 14-3-3 protein. *Transactions of the Royal Society of Tropical Medicine and Hygiene*, 108(1): 37–41. https://doi.org/10.1093/trstmh/trt097.

Zhang, M., Fu, Z., Li, C., Han, Y., Cao, X., Han, H., Liu, Y., Lu, K., Hong, Y., and Lin, J. (2015). Screening diagnostic candidates for schistosomiasis from tegument proteins of adult *Schistosoma japonicum* using an immunoproteomic approach. *PLoS Neglected Tropical Diseases*, 9(2): e0003454. https://doi.org/10.1371/journal.pntd.0003454.

Zhou, X., Chen, M., and Wu, G. (2005). "General situation." In *Asian Parasitology: Schistosomiasis in Asia (Vol.5)* edited by M. Chen, X. Zhou, K. Hirayama, 151 – 162. Chiba, Japan: AAA Committee.

Zhu, H., Yu, C., Xia, X., Dong, G., Tang, J., Fang, L., and Du, Y. (2010). Assessing the diagnostic accuracy of immunodiagnostic techniques in the diagnosis of schistosomiasis japonica: a meta-analysis. *Parasitology Research*, 107(5): 1067–1073. https://doi.org/10.1007/s00436-010-1970-3.

In: ELISA
Editor: Jose Ma. M. Angeles
ISBN: 978-1-68507-500-2

Chapter 2

USE OF AN ELISA TO ASSESS THE IMMUNOGENICITY AND PROTECTIVE POTENTIAL OF A *BABESIA BIGEMINA* RAP-1, GP45, AND 12D3 RECOMBINANT PROTEIN COCKTAIL IN CATTLE

Rebeca M. Santamaria-Espinosa, José J. Lira-Amaya, Patricia Vargas-Urióstegui, Roberto O. Castañeda-Arriola, Grecia Martínez-García, J. Antonio Álvarez-Martínez, Carmen Rojas-Martínez and Julio V. Figueroa-Millán*

Centro Nacional de Investigación Disciplinaria en Salud Animal e Inocuidad, Instituto Nacional de Investigaciones Forestales Agrícolas y Pecuarias Jiutepec, Morelos, México

* Corresponding Author's E-mail: figueroa.julio@inifap.gob.mx.

ABSTRACT

To evaluate the response of cattle immunized with a *Babesia bigemina* recombinant protein cocktail, 18 *Bos taurus* cattle randomly distributed into three groups were used. GI, the control group, was administered adjuvant on day 0, while GII and GIII were administered 100 µg of each recombinant protein (12D3, GP45, and RAP-1). Fourteen days later, a second immunization was administered to GII, while GIII animals were inoculated with an attenuated *B. bigemina* vaccine. Regarding the antibody response assessed through an indirect ELISA, the GII animals had the highest antibody response to 12D3, particularly at day 30 post-initial immunization, reaching an absorbance value of 1.4. However, the antibody response elicited from the GII cattle immunized with GP45 and RAP-1 proteins showed optical densities not greater than 0.4 absorbance units. Lower absorbance values were determined in samples from the GI and GIII cattle, even though the animals of GIII were immunized once with the cocktail containing the three recombinant proteins. At 33 days after the first inoculation, all the animals were challenged with a thawed stabilate of a *B. bigemina* field strain. The challenge led to a rise in mean rectal temperature (RT) of 39.7°C, determined on days 5-6 post-challenge (PC) in GI and GII, respectively, while no rise in RT was observed in GIII. Packed Cell Volume (PCV) values showed a slight decrease in all cattle, with the greatest decrease of 6 percentage points on day 11 PC, 17 points on day 14 PC, and 10 points on day 10 PC, in GI, GII, and GIII, respectively. *B. bigemina*-infected erythrocytes were found in peripheral blood, but none of the animals showed clinical signs associated with severe babesiosis. Immune protection could not be completely assessed as the control animals were not clinically affected by the infected blood challenge.

Keywords: *Babesia bigemina*, recombinant antigens, immunization

INTRODUCTION

Protozoan parasites of the genus *Babesia* that affect livestock and cause bovine babesiosis are mainly found in temperate and tropical regions of the world. Their distribution is limited to that of their vector, the tick *Rhipicephalus* (*Boophilus*) spp (Bock et al., 2004; de Waal and

Combrink, 2006). Tropical and subtropical regions in Mexico account for about 53% of the national territory, and more than 75% of the country's cattle herd is maintained in these regions (Bautista-Garfias & Martinez-Ibañez, 2012; Rodriguez-Vivas et al., 2017).

Important advances in the characterization of genes that can be used to encode vaccine candidate proteins have been made (Brown et al., 2006; Brown & Palmer, 1999; Wilkowsky et al., 2003; Florin-Christensen et al., 2007). Previous studies on *Babesia* have searched for antigenic proteins with immunoprotective potential, based mainly on antigenic immunodominance and conservation among *B. bigemina* isolates. The rhoptry-associated protein (RAP-1) has been shown through immunoprecipitation and immunoblotting assays to be an immunodominant antigen that is well conserved in isolates from different countries (Figueroa & Buening, 1991; Figueroa et al., 1990, 1992; McElwain et al., 1987, 1991; Machado et al., 1993, 1999; Vidoto et al., 1995).

In addition to the identification of candidate antigens based on recognition of merozoite surface proteins and the apical complex by immunized bovine serum samples (Brown et al., 1998; 2006; Figueroa et al. 1990, 1992), another alternative based on the analysis of nucleotide sequence conservation of genes encoding for antigens of interest has been proposed (Pérez et al., 2010; Ramos et al., 2012). This alternative would be of interest particularly if there was not enough parasite biological material that could be multiplied in animals or cultivated in vitro.

B. bigemina RAP-1, a protein associated with rhoptries but also present on the surface of merozoites has a relative size of 58 kDa, is expressed by a gene that is a member of a family of genes containing dimorphic regions and consists of transcripts encoding for RAP-1a, RAP-1b and RAP-1c (Mishra et al., 1991; Hotzel et al., 1996; Brown & Palmer 1999).

Previously, protection studies carried out with *Babesia bigemina* antigens such as rhoptry-associated protein 1 (RAP-1a) and merozoite surface glycoprotein GP45 showed that partial protection was induced when cattle were challenged with a homologous *B. bigemina* strain (McElwain et al., 1991). *B. bigemina* GP45 antigen is expressed on the merozoite surface and has been identified as highly immunodominant and conserved among several *B. bigemina* isolates in the Americas (McElwain et al., 1987; Figueroa et al., 1990; Mosqueda et al. 2007).

Another antigen with high vaccine potential is the 12D3 apical complex protein. According to some studies, this antigen has a high degree of conservation at the DNA level and is present in different species of *Babesia*, including *B. bigemina* (Wright et al., 1992). Immunization of cattle with the recombinant protein 12d3 from *B. bovis* has also been able to induce significant protection during a challenge with tick larvae that transmit *B. bovis* infection (Wright et al., 1992). *B. bovis* 12D3 antigen was described as a protein located in the apical complex of the *B. bovis* merozoite (Harper et al., 1996), is highly immunogenic in cattle (Alvarez et al., 2010), and encodes for a gene that presents highly conserved sequences among *B. bovis* isolates in Mexico (Perez et al., 2010). While highly conserved also among Mexican *B. bigemina* isolates (95-98% similarity of sequences), the *B. bigemina* 12D3 gene sequence only has 68% sequence identity with the *B. bovis* 12D3 gene (Lira-Amaya et al., 2017).

The conservation of RAP-1 epitopes or gene sequences among several *B. bigemina* strains support continued testing of RAP-1, GP45, and 12D3 as a vaccine component. However, the immunoprotective capacity of these antigens in cattle has not been tested in combined form as a cocktail. The objective of the present study was to evaluate the immunogenicity and immunoprotective potential of a cocktail of *Babesia bigemina* recombinant proteins (rRAP-1, rGP45, and r12D3), through clinical monitoring of cattle that were immunized and challenged under controlled experimental conditions.

MATERIALS AND METHODS

Recombinant Proteins

Recombinant proteins were obtained by extracting genomic DNA from a Mexican strain of *Babesia bigemina* and PCR amplifying the truncated coding regions of *rap-1α*, *gp45* and *12d3*, with the primer sets *rap-1* NF 5´acggcggctaagttcttc-3´/ *rap-1* NR 5´-agaagtctacaaagtctacgc-3´; *gp45* NF 5´-actagtcgcacaagtgccc-3´/ *gp45* NR 5´ctcacgacgaaaacgtaaa-3´; and *12d3* NF 5´-atgatgctcgctacgtttt-3´/ *12d3* NR 5´-gctggaacaactttgttaatttagaatgatc-3´, respectively (Santamaria et al., 2020). The amplified regions of the *rap-1α*, *gp45*, and *12d3* genes were identified in agarose gels (not shown) and corresponded to the expected sizes of 840 bp for *rap-1α*, 963 bp for *gp45*, and 798 bp for *12d3*, according to the nucleotide sequences that have been reported for *B. bigemina* (GenBank accession numbers M60878, AF298630, and A23051, respectively).

Subsequently, DNA amplicons were cloned into the pBad/TOPO® Thio Fusion™ expression vector kit (Invitrogen, USA). Expression of the protein was induced using IPTG in bacterial cultures, and colonies containing the gene insert of interest were selected, DNA sequenced, and subjected to recombinant protein induction and purification, essentially as previously reported (Borgonio et al., 2008). Lastly, recombinant proteins were purified by affinity chromatography using columns with NI-NTA resin. The purification quality was analyzed through electrophoresis on 12% SDS-PAGE polyacrylamide gels, and the protein concentration was determined by fluorometry (Qubit, Invitrogen). At the end of the run, the proteins were dry transferred in an iBlott™ assembly to a nitrocellulose (NC) membrane. The nitrocellulose membrane was exposed to a bovine serum containing antibodies to the recombinant proteins, essentially as described previously (Borgonio et al., 2008).

Experimental Animals

A total of 18 one-year-old *Bos taurus* steers, native from a *Rhipicephalus (Boophilus) microplus*-free zone (and therefore free from bovine babesiosis) were selected and randomly distributed into three groups (GI, GII, and GIII). GI steers, which served as a control group throughout the experiment, were inoculated on days 0 and 14 with 1 ml of Montanide 75 adjuvant. In the first inoculation to GII and GIII animals, steers received 100 μg of each of the rRAP-1, rGP45, and r12D3 proteins, diluted in 500 μl of phosphate-buffered saline (PBS) and mixed in a 1:1 ratio (vol/vol) with adjuvant, such that the final total volume was 1 ml/animal. After 14 days, a second inoculation was administered to GII steers, with the same dose of recombinant proteins (final total volume of 1 ml/animal), while GIII steers were inoculated with an attenuated vaccine strain, consisting of a thawed stabilate of 1 x 10^8 *B. bigemina*-infected erythrocytes (EI), derived from in vitro culture (Alvarez et al., 2004; Rojas-Martinez et al., 2018). The intramuscular route of administration was used for inoculation of the live attenuated vaccine. The purpose of inoculating the GIII animals firstly with the protein cocktail and then, two weeks later, with the live attenuated vaccine was to try to improve the effectiveness of the protection that the live attenuated vaccine provides (usually 80%). Subsequently, 33 days after the first inoculation (day 0 post-challenge), the three groups of steers (GI, GII, and GIII) were experimentally challenged by inoculating a thawed stabilate containing 1 x 10^8 *B. bigemina*-infected erythrocytes of a field strain that had been isolated in Chiapas State, Mexico (Lira-Amaya et al., 2017).

The animals were handled in accordance with the guidelines for good management and animal welfare established through NOM-062-ZOO-1999 (Technical specifications for production, care, and use of laboratory animals) in Mexico (http://www.senasica.gob.mx/?doc=743).

Clinical Monitoring and Sampling

Clinical monitoring of experimental steers was carried out during both, the vaccination, and the challenge phase, by recording the individual rectal temperature (RT) and packed cell volume (PCV) values every day. At the same time, blood samples were collected, and Giemsa-stained blood smears were microscopically analyzed to detect the presence of *B. bigemina* and to determine the percentage of parasitized erythrocytes (PPE). In addition, plasma samples were obtained from the steers every 15 days (totaling 7 sampling dates), using EDTA (ethylenediaminetetraacetic acid) as the anticoagulant. The presence of specific antibodies in blood plasma was determined using the indirect fluorescent antibody test (IFAT) and an indirect ELISA with the *B. bigemina* recombinant proteins as the antigen (Santamaria et al., 2020). In parallel, body condition and clinical signs that might suggest the presence of overt disease were evaluated (Alvarez et al., 2004, 2010; Bautista-Garfias & Martinez-Ibañez, 2012).

Statistical Analysis

Statistical analysis on differences in clinical protection against the disease between the vaccinated experimental and control groups was performed using Fisher's exact test. Statistically significant differences were considered to exist when p-values were found to be less than or equal to 0.05 ($P < 0.05$).

Results and Discussion

During the vaccination phase, there were no adverse reactions among the GII and GIII animals. Regarding the antibody response assessed through the indirect ELISA, Figure 1 shows that the GII steers had the highest antibody response to r12D3 antigen, particularly at day 30 post-initial immunization, reaching an absorbance value of 1.4, which subsequently decreased at the next sampling times. However, the antibody response elicited to rGP45 and rRAP-1 antigens by the GII steers showed optical densities not greater than 0.4 absorbance units (Figure 1). Lower absorbance values were also determined in samples from the GI and GIII steers, even though the GIII steers were immunized once with the cocktail containing the three recombinant proteins (r12D3, rGP45, and rRAP-1).

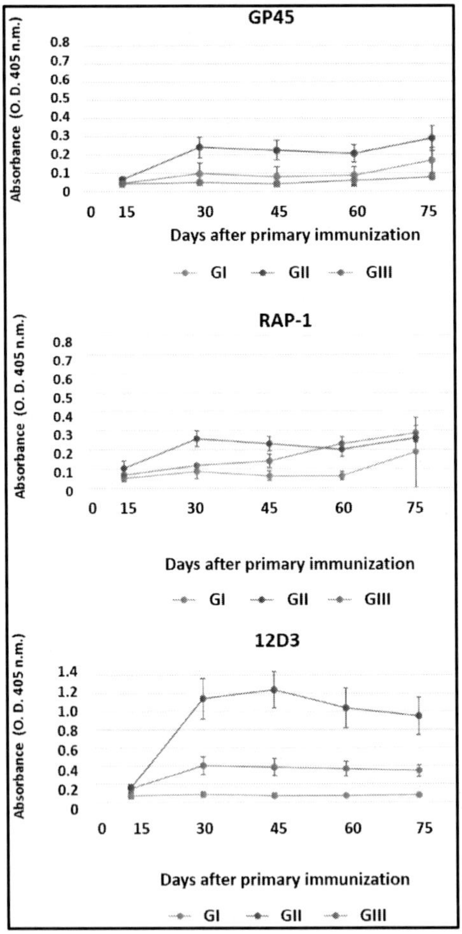

Figure 1. Cattle antibody levels in response to immunization with *B. bigemina* recombinant proteins rGP45, rRAP-1, and r12D3 as assessed by the iELISA.

On the other hand, analysis of the serum samples by the IFAT showed that all the GII steers seroconverted after the second immunization, thus indicating that the antibodies recognized authentic native *B. bigemina* antigens (see example in Figure 2), and all the animals were determined positive by the IFAT at the end of the experiment (day 75 after primary immunization).

Previous studies reporting on the use of native or recombinant parasite proteins as immunizing antigens to induce protection against bovine babesiosis (McElwain et al., 1991; Wright et al., 1992; Machado et al., 1999; Alvarez et al., 2010) showed that specific antibodies recognizing either the native or the recombinant parasite antigens can be detected by using various immunological tests such as the IFAT, ELISA, immunoblotting and immunoprecipitation tests. However, despite induction of humoral response, such as that measured using the ELISA and fluorescent antibody assays among immunized cattle, the presence of antibodies does not necessarily give rise to immune protection. This has been shown in reports of immunized cattle challenged with virulent strains in which adequate protection was not achieved (Fish et al., 2008; Alvarez et al., 2010; Reyes-Sandoval et al., 2016). Previous studies in which the live attenuated vaccine against bovine babesiosis was evaluated showed the induced antibody kinetics as assessed by the IFAT. The maximum antibody titers for *Babesia bovis* and *Babesia bigemina* occurred on day 20 after immunization (Cantó-Alarcón et al., 1999; Bautista-Garfias & Martinez-Ibañez, 2012).

Regarding the monitoring of clinical signs during the immunization phase of this study, fever was not detected in any of the experimental steers. The average RT values were 38.8°C and 38.9°C, for the GII and GIII steers, respectively. No clinical alterations associated with the disease were observed during this phase, even though the GIII steers received a second immunization with a live attenuated *Babesia bigemina* vaccine. Throughout the vaccination phase, the PCV value was determined to be within normal values, with mean PCV of 32.17% and 30.58% for the GII and GIII steers, respectively, even though the GIII steers received a 1 x 10^8 IE dose of the thawed *B. bigemina* vaccine strain as the second inoculation. The presence of vaccine parasites in the GIII steers was determined by optical microscopy analysis of Giemsa-stained smears, in which *Babesia bigemina* parasites were identified in 5 out of 6 (83.3%) of the GIII steers. This, therefore indicated that the vaccine parasites had become established in the cattle inoculated with the live attenuated vaccine (Table 1). The estimated parasitemia was <0.01% and only occurred over days 6-10 post-inoculation of the *B. bigemina*-infected erythrocyte stabilate.

During the challenge phase, an increase in the mean RT value (up to 39.7°C) was determined on day 5 post-challenge (PC) in the GI steers and up to 39.6°C during days 5 and 6 PC in the GII steers, whereas no rise in the mean RT value was observed in the GIII animals (Figure 3).

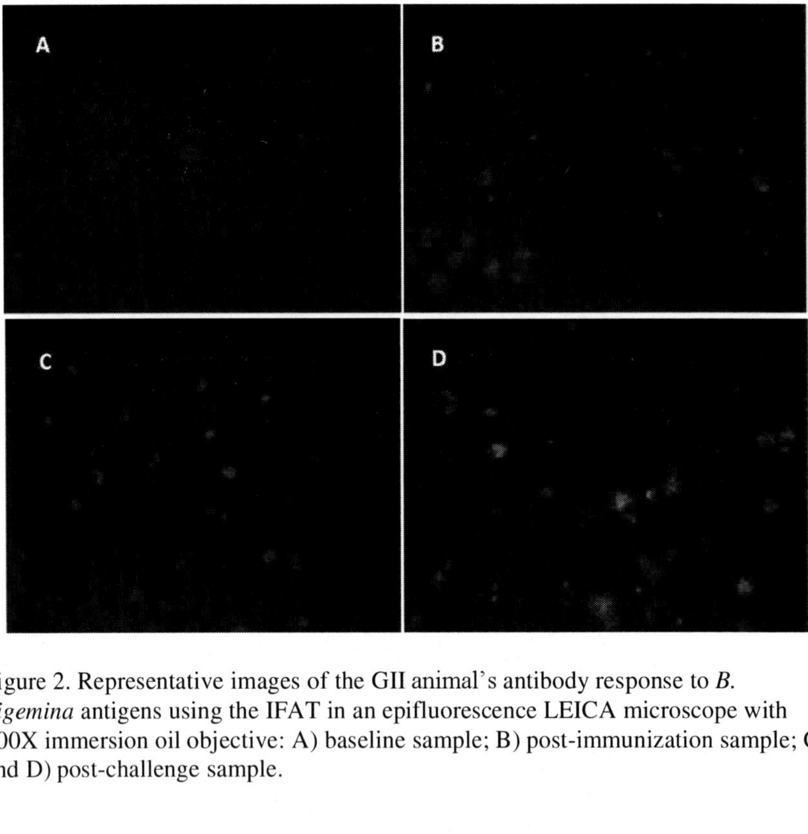

Figure 2. Representative images of the GII animal's antibody response to *B. bigemina* antigens using the IFAT in an epifluorescence LEICA microscope with 100X immersion oil objective: A) baseline sample; B) post-immunization sample; C and D) post-challenge sample.

Fever was determined to be always higher in all the control group steers, reaching up to 41°C (Table 2). These findings relate to the pathophysiology of the disease, in which endogenous pyrogens are produced, especially IL-1, which is associated with the immunopathology of *Babesia* infection (Brown & Palmer, 1999). A marked decrease in the mean PCV value (22.8%) was determined on day 11 PC in the GIII steers, as compared to the baseline mean PCV (30.6%) (Table 2). In the GI and GII steers, a decrease in PCV was also identified, with mean PCV values of 24.8% on day 11 PC and 25.3% on day 13 PC, respectively (Figure 4). In previous studies using live attenuated vaccines to protect against bovine babesiosis, maximum decreases in PCV of 21.4% and 33.4% were found when using frozen vaccine doses of 1 x 10^8 IE for *B. bovis* and 1 x 10^9 IE for *B. bigemina* (Cantó-Alarcón et al., 1999), and when using a combined dose of fresh vaccine containing 1 x 10^7 IE each of *B. bovis* and *B. bigemina* (Cantó-Alarcón et al., 2003). In comparison with the results from other studies carried out with attenuated vaccines, the decrease in PCV found in the present study was not considered severe, since in most other cases the challenge was conducted in the field and in the presence of the tick vector (Figueroa-Millán et al., 1998; Cantó-Alarcón et al., 1999; Cantó-Alarcón et al., 2003).

Table 1. Identification of *B. bigemina*-infected erythrocytes in GIII experimental animals vaccinated with a frozen stabilate containing 1x10^8 attenuated parasites

ID / DPI	6	7	8	9	10	11
1	-	-	-	-	-	-
2	-	-	+	+	-	-
3	+	+	+	+	-	-
4	-	+	-	-	-	-
5	-	-	+	-	-	-
6	-	+	+	+	+	-

ID = Animal identification. DPI= Days post-immunization.

Positive animals showed between 1-2 merozoite- or trophozoite-infected erythrocytes in a 10 min microscopic search, covering 100 microscopic fields with at least 200 erythrocytes each.

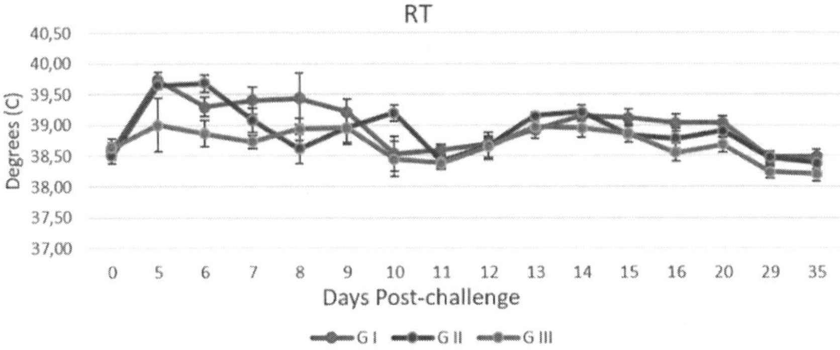

Figure 3. Mean Rectal Temperature value (RT) in experimental groups during the challenge phase. Each point represents the mean + SE, for each group.

Regarding the PPE values, a PPE of 0.22 was estimated in the GI steers on day 9 PC, whereas PPEs of 0.10 on day 7 PC and 0.17 on day 9 PC were detected in the GII and GIII steers, respectively (Table 2). None of the animals showed clinical signs associated with severe bovine babesiosis at the challenge and therefore did not need babesiacide treatment.

The results obtained during the challenge phase in this study were like those reported previously (Reyes-Sandoval et al., 2016), in which the immune protective capacity and efficacy of a cocktail of *Babesia bovis* recombinant proteins (merozoite surface antigens MSA-1, MSA-2c, and 12D3) were evaluated as candidates for vaccines against bovine babesiosis. However, unlike the present study, the challenge was performed in the field and the group of cattle that was inoculated with the recombinant protein cocktail did not show any protection, as the immunized animals showed severe clinical signs associated with babesiosis and had to be treated with a babesiacide. In the present study, the data did not show independence between the vaccination and the clinical disease protection through nonparametric analysis, such as Fisher's exact test ($P < 0.05$). This probably occurred because of the limited virulence of the field strain used, along with the comfortable environment in which the experimental animals were kept.

Thus, no babesiacide treatment was required in the present study, unlike the required treatment done in previous studies (Reyes-Sandoval et al., 2016).

Table 2. Clinical variables determined in experimental cattle after challenge with a frozen stabilate containing 1 x 10^8 *B. bigemina*-infected erythrocytes of a field isolate

Variable	GI	GII	GIII
Incubation period (Days PC)	5	5	7
Days with Fever (>39.5ºC)	2	2	0
Maximum Individual RT (ºC)	41.0	41.3	39.7
Animals with Fever (>39.5ºC)	6	5	2
Minimum PCV (mean / group)	24.8	25.3	22.8
Minimum Individual PCV	21	21	13
Mean Percent Parasitized Erythrocytes	0.22	0.1	0.17

PC = Post Challenge; RT = Rectal temperature; PCV = Packed Cell Volume.
PPE = Percent of Parasitized Erythrocytes.

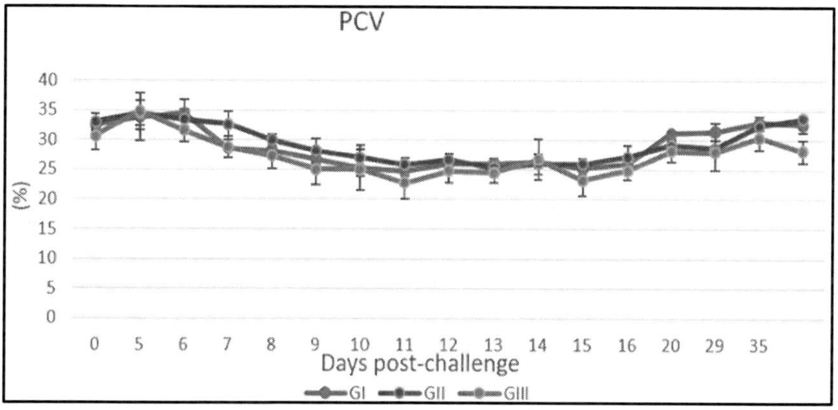

Figure 4. Mean Packed Cell Volumen value (PCV %) in experimental groups during the challenge phase. Each point represents the mean + SE, for each group.

In contrast, in a study conducted in Israel, in which the efficacy of a live attenuated *Babesia bovis* vaccine, a vaccine based on in vitro culture exoantigens, and a subunit vaccine containing recombinant rhoptry-associated protein (RAP)-1 were compared as immunogens in cattle challenged under controlled conditions, it was concluded that while both the live parasites and exoantigens induced protection in immunized cattle, the use of the recombinant protein only generated partial protection against challenge (Fish et al., 2008). Furthermore, in a study in which the immunoaffinity-purified native *B. bigemina* RAP-1a, GP45, and GP55 proteins were used as individual immunogens in cattle, only partial protection was found when immunized cattle were challenged with a homologous *B. bigemina* strain (McElwain et al., 1991).

CONCLUSION

The iELISA reported in this study can adequately be used to assess the antibody response in cattle that have been immunized with a cocktail of *B. bigemina* antigens including recombinant proteins RAP-1α, GP45, and 12D3. However, from the results obtained in this study, it is not possible to conclude whether the cocktail of *Babesia bigemina* recombinant proteins used as immunogens or the protein cocktail plus the frozen attenuated vaccine induced an immunoprotective response. Most likely, the cryopreserved blood did not contain the most appropriate dose of IE for an adequate challenge. Nonetheless, the clinical data obtained after monitoring the immunization of cattle indicates that using both, the recombinant proteins and the combination of recombinant proteins and live attenuated vaccine, were harmless for cattle, as no significant changes in the variables evaluated in this study were identified as compared to the non-immunized steers of the control group. The level of protection conferred by recombinant proteins in animals immunized and challenged under field conditions with the tick vector transmitting the *Babesia* parasites remains to be assessed.

ACKNOWLEDGMENTS

This study was partially funded by CONACYT, Mexico, Problemas Nacionales 2015, Project No. 1336; SEP-CONACYT CB2017-2018 Project No. A1-S-43508; and INIFAP, Project No. 1672534936.

REFERENCES

Alvarez, J. A., Ramos, J. A., Rojas, E. E., Mosqueda, J. J., Vega, C. A., Olvera, A. M., Figueroa, J. V. and Cantó, G. J. (2004). Field challenge of cattle vaccinated with a combined *Babesia bovis* and *Babesia bigemina* frozen immunogen. *Annals of the New York Academy of Sciences*, 1026: 277-283.

Alvarez, J. A., Lopez, U., Rojas, C., Borgonio, V. M., Sanchez, V., Castaneda, R. Vargas, P. and Figueroa, J. V. (2010). Immunization of *Bos taurus* steers with *Babesia bovis* recombinant antigens MSA-1, MSA-2c and 12D3. *Transboundary and Emerging Diseases,* 57: 87-90.

Bautista-Garfias, C. R. and Martinez-Ibañez, F. (2012). Experiences on the control of cattle tick *Rhipicephalus (Boophilus) microplus* in Mexico. In: Moges Woldemeskel (ed) Ticks: Disease, management, and control. *Nova Science Publishers, Inc.*, pp. 205-216.

Bock, R. E., Jackson, L., de Vos, A. J. and Jorgensen, W. (2004). Babesiosis of cattle. *Parasitology*, 4: 247-269.

Borgonio, V. M., Mosqueda, J., Genis, A. D., Falcon, A., Alvarez, J. A., Camacho, M. and Figueroa, J. V. (2008). *msa-1* and *msa-2c* gene analysis and common epitopes assessment in Mexican *Babesia bovis* isolates. *Annals of the New York Academy of Sciences*, 1149: 145-148.

Brown, W. C. and Palmer, G. H. (1999). Designing blood stage vaccines against *Babesia bovis* and *Babesia bigemina. Parasitology Today*, 15: 275-281.

Brown, W. C., McElwain, T. F., Hotzel, I., Suarez, C. E. and Palmer, G. H. (1998). Helper T cell epitopes encoded by the *Babesia bigemina rap-1* gene family in the constant and variant domains are conserved among parasite strains. *Infection & Immunity,* 66: 1561-1569.

Brown, W. C., Norimine, J., Goff, W. L., Suarez, C. E. and McElwain, T. F. (2006). Prospects for recombinant vaccines against *Babesia bovis* and related parasites. *Parasite Immunology*, 28: 315-327.

Cantó-Alarcón, G. J., Figueroa-Millán, J. V., Ramos-Aragón, J. A., Mosqueda-Gualito, J. J. and Vega y Murguía, C. A. (1999). Evaluación de la capacidad protectora de un inmunógeno fresco combinado de *Babesia bigemina* y *Babesia bovis. Revista Veterinaria Mexico,* 30: 215-219. [Evaluation of the protective capacity of a combined fresh immunogen of *Babesia bigemina* and *Babesia bovis. Veterinary Mexico Magazine,* 30: 215-219].

Cantó-Alarcón, G. J., Rojas-Ramírez, E. E., Alvarez-Martínez, J. A., Ramos-Aragón, J. A., Mosqueda-Gualito, J. J., Vega y Murguía, C. A. and Figueroa-Millán, J. V. (2003). Protection against bovine babesiosis with a mixed *in vitro* culture derived *B. bovis* and *B. bigemina* vaccine under a field challenge. II. Immunization in an endemic area. *Tecnica Pecuaria Mexico*, 4: 307-315.

de Waal, D. T. and Combrink, M. P. (2006). Live vaccines against bovine babesiosis. *Veterinary Parasitology*, 138: 88-96.

Fish, L., Leibovich, B., Krigel, Y., McElwain, T. F. and Shkap, V. (2008). Vaccination of cattle against *B. bovis* infection with live attenuated parasites and non-viable immunogens. *Vaccine*, 6: 29-33.

Figueroa, J. V. and Buening, G. M. (1991). In vitro inhibition of multiplication of *Babesia bigemina* by using monoclonal antibodies. *Journal of Clinical Microbiology*, 29: 997-1003.

Figueroa, J. V., Buening, G. M., Kinden, D. A. and Green, T. J. (1990). Identification of common surface antigens among *B. bigemina* isolates by using monoclonal antibodies. *Parasitology,* 100: 161-175.

Figueroa, J. V., Alvarez, J. A., Buening, G. M., Canto, G. J., Hernandez, R., Monroy, M., Ramos, J. A. and Vega, C. A. (1992). Antibody response to *Babesia bigemina* infection in calves measured by

ELISA and immunoblotting techniques. *Revista LatinoAmericana de Microbiologia*, 34(4): 267-273.

Figueroa-Millán, J. V., Cantó-Alarcón, G. J., Alvarez-Martínez, J. A., Lona-Gálvez, R., Ramos-Aragón, J. A. and Vega-Murguía, C. A. (1998). Capacidad protectora en bovinos de una cepa de *Babesia bigemina* derivada del cultivo *in vitro*. *Tecnica Pecuaria en Mexico,* 36: 95-100. [Protective capacity in cattle of a strain of *Babesia bigemina* derived from in vitro culture. *Livestock Technique in Mexico,* 36: 95-100.]

Florin-Christensen, M., Schnittger, L., Dominguez, M., Mesplet, M., Rodriguez, A., Ferreri, L., Asenzo, G., Wilkowsky, S., Farber, M., Echaide, I. and Suarez, C. (2007). Search for *Babesia bovis* vaccine candidates. *Parasitologia*, 49(S1): 9-12.

Harper, G. S., Hibbs, A. R., East, I. J., Waltisbuhl, D. J., Jorgensen, W. K. and Riddles, P. W. (1996). *Babesia bovis*: biosynthesis and localization of 12D3 antigen in bovine erythrocytes. *International Journal for Parasitology,* 26: 1255-1262.

Hotzel, I., Suarez, C. E., McElwain, T. F. and Palmer, G. H. (1997). Genetic variation in the dimorphic regions of *rap-1* genes and *rap-1* loci of *Babesia bigemina*. *Molecular and Biochemical Parasitology*, 90: 479 - 489.

Lira-Amaya, J. J., Vargas-Uriostegui, P., Pérez de la Rosa, J. D., Rojas-Martínez, M. C., Álvarez-Martínez, J. A. and Figueroa-Millán, J. V. (2017). Conservación de la secuencia del gen ortólogo del antígeno 12D3 de *Babesia bigemina* en aislados mexicanos. *Memoria Reunión Nacional de Investigación Pecuaria.* 3(1): 594-595. [Sequence conservation of the orthologous gene for 12Dd antigen of *Babesia bigemina* in Mexican isolates. *Proceedings National Livestock Research Meeting, Mexico,* 3(1): 594-595]. http://reunion esnacionales.inifap.gob.mx/RNIP/_MEMORIAS/2017.pdf.

Machado, R. Z., McElwain, T. F., Suarez, C., Hines, S. A. and Palmer, G. H. (1993). *Babesia bigemina*: Isolation and characterization of merozoite rhoptries. *Experimental Parasitology*, 77(3): 315-325.

Machado, R. Z., McElwain, T. F., Pancracio, H. P., Freschi, C. R. and Palmer, G. H. (1999). *Babesia bigemina*: Immnunization with purified rhoptries induces protection against acute parasitemia. *Experimental Parasitology*, 93(2): 105-108.

McElwain, T. F., Perryman, L. E., Davis, W. C. and McGuire, T. C. (1987). Antibodies define multiple proteins with epitopes exposed on the surface of live *Babesia bigemina* merozoites. *Journal of Immunology*, 138: 2298-2304.

McElwain, T. F., Perryman, L. E., Musoke, A. J. and McGuire, T. (1991). Molecular characterization and immunogenicity of neutralization sensitive *Babesia bigemina* merozoite surface proteins. *Molecular and Biochemical Parasitology*, 47: 213-222.

Mishra, V. S., McElwain, T. F., Dame, J. B. and Stephens, E. B. (1992). Isolation, sequence, and differential expression of the p58 gene family of *Babesia bigemina*. *Molecular and Biochemical Parasitology*, 53: 149-158.

Mosqueda, J., Figueroa, J. V., Alvarez, A., Bautista, R., Falcon, A., Ramos, A., Canto, G. J. and Vega, C. A. (2007). Advances in the development of molecular tools for the control of bovine babesiosis in Mexico. *Parassitologia*, 49(S1): 19-22.

Perez, J., Perez, J. J., Vargas, P., Alvarez, J. A., Rojas, C. and Figueroa, J. V. (2010). Sequence conservation of the 12D3 gene in Mexican isolates of *Babesia bovis*. *Transboundary and Emerging Diseases*, 57: 57-60.

Ramos, C. A. N., Araujo, F. R., Alves, L. C., Souza, I. I. F., Gueder, D. S. and Soares, C. O. (2012). Genetic conservation of potentially immunogenic proteins among Brazilian isolates of *Babesia bovis*. *Veterinary Parasitology*, 187: 548-552.

Reyes-Sandoval, R. M., Bautista-Garfias, C. R., Castaneda-Rios, R. O., Vargas-Uriostegui, P., Alvarez-Martinez, J. A., Rojas-Martinez, C., Mejia-Estrada, F. and Figueroa-Millán, J. V. (2016). Babesiosis: Field assessment of protection in cattle immunized with a mixture of *Babesia bovis* recombinant proteins. *Quehacer Cientifico en Chiapas*, 11: 36-46.

Rodriguez-Vivas, R. I., Grisi, L., Perez de Leon, A. A., Silva-Villela, H., Torres-Acosta, J. F. J., Fragoso-Sanchez, H., Romero-Salas, D., Rosario-Cruz, R., Saldiernah, F. and García-Carrasco, D. (2017). Potential economic impact assessment for cattle parasites in Mexico. Review. *Revista Mexicana de Ciencias Pecuarias*, 8: 61-74.

Rojas-Martínez, C., Rodríguez-Vivas, R. I., Figueroa-Millán, J. V., Bautista-Garfias, C. R., Castañeda-Arriola, R. O., Lira-Amaya, J. J., Vargas-Urióstegui, P., Ojeda-Carrasco, J. J. and Alvarez-Martínez, J. A. (2018). Bovine babesiosis: Cattle protected in the field with a frozen vaccine containing *Babesia bovis* and *Babesia bigemina* cultured in vitro with a serum-free medium. *Parasitology International*, 67: 190-195.

Santamaria, R. M., Lira, J. J., Vargas, P., Alvarez, J. A., Rojas, C., Figueroa, J. V. (2020). Validation of an indirect ELISA using recombinant proteins as antigen to identify animals exposed to *Babesia bigemina*. *Transboundary Emerging Diseases*, 67(S2): 201-207.

Vidoto, O., McElwain, T. F., Machado, R. Z., Perryman, L., Suarez, C. and Palmer, G. H. (1995). *Babesia bigemina*: Identification of B cell epitopes associated with parasitized erythrocytes. *Experimental Parasitology*, 81(4): 491-500.

Wilkowsky, S. E., Farber, M., Echaide, I., Torioni de Echaide, S., Zamorano, P. I., Dominguez, M., Suarez, C. E. and Florin-Christensen, M. (2003). *Babesia bovis* merozoite surface protein-2c (MSA-2c) contains highly immunogenic, conserved B-cell epitopes that elicit neutralization-sensitive antibodies in cattle. *Molecular and Biochememical Parasitology*, 127: 133-141.

Wright, I. G., Casu, R., Commins, M. A., Dalrymple, B. P., Gale, K. R., Goodger, B. V., Riddles, P. W., Waltisbuhl, D. J., Abetz, I., Berrie, D. A., Bowles, Y., Dimmock, C., Hayes, T., Kalnins, H., Leatch, G., McCrae, R., Montague, P. E., Nisbet, I. T., Parrodi, F., Peters, J. M., Scheiwe, P. C., Smith, W., Rode-Bramanis, K. and White, M. A. (1992). The development of a recombinant *Babesia* vaccine. *Veterinary Parasitology*, 44: 3-13.

In: ELISA
Editor: Jose Ma. M. Angeles
ISBN: 978-1-68507-500-2

Chapter 3

ELISA TEST USING *BABESIA* SP. RECOMBINANT ANTIGENS FOR EPIDEMIOLOGICAL DIAGNOSIS OF BOVINE BABESIOSIS IN MEXICO

Julio V. Figueroa-Millán**, Roberto O. Castañeda-Ríos, José J. Lira-Amaya, Patricia Vargas-Urióstegui, J. Antonio Álvarez-Martínez and Carmen Rojas-Martínez

Centro Nacional de Investigación Disciplinaria en Salud Animal e Inocuidad, Instituto Nacional de Investigaciones Forestales Agrícolas y Pecuarias Jiutepec, Morelos, México

ABSTRACT

The objective of this study was to evaluate an ELISA method for the epidemiological diagnosis of bovine babesiosis using the recombinant protein RAP-1 of *Babesia bigemina* and MSA-1 of

* Corresponding Author's E-mail: figueroa.julio@inifap.gob.mx.

Babesia bovis as antigens. Diagnostic sensitivity and specificity rate estimates for the ELISA test with RAP-1 were 90% and 93.75% respectively. The diagnostic sensitivity and specificity rates of the ELISA with MSA-1 were 79% and 95%, respectively. The analysis performed with the ELISA test in 103 serum samples of a bovine herd from Paso del Toro, Veracruz yielded the following results: 23 serum samples (22.34%) were negative for both tests, 14 samples (13.59%) were positive in the ELISA-RAP-1 (*B. bigemina*), 29 samples (28.15%) were positive in the ELISA-MSA-1 (*B. bovis*), while 37 samples (35.92%) were identified as positive in both tests. Thus, the prevalence of antibodies to *Babesia* sp in the cattle herd determined using ELISA with recombinant antigen is estimated at 77.66%, slightly higher than the value obtained with the indirect immunofluorescence test (71%), considered the standard test to estimate the seroprevalence of bovine babesiosis in Mexico.

Keywords: Bovine babesiosis, *Babesia bigemina*, *Babesia bovis*, ELISA

INTRODUCTION

Bovine babesiosis, caused by intraerythrocytic parasites of the genus *Babesia*, is a tick-borne disease widely distributed in tropical and subtropical regions where the vector tick, *Rhipicephalus* spp., is endemic (McCosker, 1981; Bock et al., 2004; Maharana et al., 2016). *Babesia bigemina* and *B. bovis* are the etiological agents of bovine babesiosis in Mexico which primarily affect beef cattle, double purpose cattle, and dairy cattle grazing in tropical areas of the country (Alvarez et al., 2019; 2020). Routine clinical diagnosis for acute cases of bovine babesiosis is based on microscopic detection of intra-erythrocytic parasites in blood smears (Bose et al. 1995; Maharana et al., 2016; Alvarez et al., 2019). The test is relatively laborious when many blood samples need to be simultaneously evaluated and, in addition, the detection limit for optical microscopy is not always sensitive enough in cases with chronic or persistent infection (Bose et al., 1995; Figueroa & Buening, 1995; Alvarez et al., 2019). An alternative for diagnosing previous exposure to *Babesia* spp in livestock is the indirect identification of anti-*Babesia* antibodies by serological methods (Bose et al., 1995; Boonchit, et al., 2004; Goff et al., 2003; 2008) such as the Indirect Fluorescent Antibody Test (IFAT) or an ELISA, or directly through the detection of nucleic acids using highly analytically sensitive but relatively costly methods in terms of application, such as the PCR test (Figueroa et al., 1993; Figueroa & Buening, 1995; Martinez-Garcia et al., 2020). The IFAT has been extensively used to estimate seroprevalence rates of bovine babesiosis (Alvarez et al., 2019). In Mexico, prevalence rates ranging from 5% to 89% have been identified (Álvarez-Martinez & Cantó-Alarcón 1985; Alvarez et al., 2019). Given the high subjectivity of the IFAT, the need for more objective, versatile, and reproducible serological evidence has been demanded in Mexico which, in addition to having high sensitivity and epidemiological specificity, lends himself to a semi-automation scheme that allows the analysis of a high number of samples at once.

ANALYTICAL FRAMEWORK

Literature Review

In the past, immunoprecipitation, and immunoblotting assays have previously demonstrated that the Ropthry-Associated Protein (RAP-1) of *B. bigemina* is an immunodominant antigen highly conserved in parasite isolates of different geographical origins (Figueroa et al., 1990; McElwain et al, 1991; Figueroa et al., 1992; Figueroa et al., 1998). Similarly, the Merozoite Surface Antigen (MSA-1) is an immunodominant antigen with high conservation in different isolates of *B. bovis* in Mexico and other countries (Hines et al., 1989; Goff et al., 1988; Florin-Christensen et al., 2002; Figueroa et al., 1998; Borgonio et al., 2008). Currently, the IFAT is the gold standard assay and has been used widely in Mexico to carry out bovine babesiosis serosurveillance studies. However, the cross-reaction between *B. bovis* and *B. bigemina* is a major drawback for species-specific diagnosis (Alvarez et al. 2019). The results are influenced by the operator's subjective judgment, while the IFAT provides adequate sensitivity and it is easy to perform, its reduced specificity makes it non-viable in some laboratories and is not routinely followed nowadays. These are the main drawbacks that impede its wide use in diagnostic laboratories established around the country. As a preliminary step in attempting to change this situation in Mexico, the present study aimed to assess two recombinant proteins: *B. bigemina* recombinant protein RAP-1 and *B. bovis* recombinant MSA-1 as antigens in an indirect ELISA which could eventually be useful to conduct epidemiological diagnosis studies of bovine babesiosis in Mexico.

METHOD

Recombinant Protein

The RAP-1 and MSA-1 antigens were obtained by first PCR amplifying the encoding sequences of the *rap-1* and *msa-1* genes, using genomic DNA from *B. bigemina* and *B. bovis* as a specific mold and appropriate oligonucleotides (Borgonio et al., 2008; Alvarez et al., 2010; Santamaria et al., 2020) using the gel-purified DNA amplicons for cloning and expression in the pBAD/TOPO Thio-expression system (Invitrogen). The recombinant version of the *B. bigemina* RAP-1 and *B. bovis* MSA-1 antigens were purified by affinity chromatography using columns with NI-NTA resin and the purified protein's quality was analyzed by electrophoresis in SDS polyacrylamide gels (Alvarez et al., 2010; Santamaria et al. 2020).

Control Sera

At least 30 serum samples from animals considered truly negative (no infection) were used for the *B. bovis* iELISA, whereas 80 negative samples were used for the *B. bigemina* iELISA test. Thirty bovine sera from experimentally vaccinated or infected with strains of *Babesia bigemina* or *B. bovis* were utilized. These sera were considered as truly positive samples. Samples obtained from both groups of animals were used for the implementation of the iELISA and the determination of the diagnostic sensitivity and specificity rates.

Enzyme-Linked Immunoabsorbent Assay (ELISA)

Briefly, ELISA plates of 96 wells were used; 100 μl of the rRAP-1 or rMSA-1 protein were placed at a concentration of 1 mg/mL in carbonate solution (pH 9.6) and incubated at 4°C overnight. For the blocking step, 300 μL/well of the blocking solution (PBS with 5% skim milk) were used and plates were incubated at 37°C for 1 hr. Washing steps were performed using a PBS solution (pH 7.2). The bovine serum samples were diluted at 1:100 in PBS and incubated at 37°C for 30 min. Then, rabbit anti-IgG conjugated with alkaline phosphatase at a 1:10,000 dilution was added, and plates were incubated at 37°C for 30 min. Subsequently, p-nitrophenyl phosphate substrate-chromogen was added. Plate absorbance reading was done after incubating at 37°C for 30 min at an optical density of 405 nm. For positive/negative sample result classification, the mean absorbance values obtained for the negative samples were determined and the equivalent of two standard deviations from the absorbance average was added to it to establish the threshold.

Indirect Fluorescent Antibody Test (IFAT)

Infected erythrocytes derived from the *in vitro* cultivation of *B. bigemina* (Alvarez et al., 2004; Rojas-Martínez et al., 2018) or *B. bovis* (Alvarez et al., 2004; Rojas et al., 2004) with parasitemias of >5% were utilized to prepare thin blood smears on microscope slides which were maintained at -20°C until used as antigen for the IFAT. When performing the technique, the blood smears were desiccated at 37°C for 15 min; each serum sample was diluted at 1:80 with PBS and placed in circles previously marked with fatty pencil on the blood smears and incubated at 37°C for 30 min. Slides were washed with PBS and incubated with the rabbit anti-bovine IgG conjugated with fluorescein isothiocyanate. Slides were then incubated at 37°C for 30 min and washed with PBS as before. Finally, slides were washed in PBS as before with a final wash in distilled water, dried, and observed under an epifluorescence microscope at 100X.

Data Analysis

Contingency tables (2 x 2) were used to determine the sensitivity and specificity of the ELISA. In addition, the kappa (k) value between the IFAT and ELISA techniques was determined (Landis and Koch, 1977; Simoes et al. 1995).

Evaluation of the iELISA Tests with rRAP-1 and rMSA-1 Antigens in Samples of Animals from an Endemic Area of Bovine Babesiosis

A total of 103 bovine serum samples were analyzed from a representative sampling of a bovine herd located at "La Posta" experimental station in Veracruz, Mexico, an endemic area of bovine babesiosis where *Babesia sp.* prevalences of up to 80% have previously been determined (Álvarez-Martinez & Cantó-Alarcón, 1985; Alvarez et al. 2019; Rojas-Martinez et al. 2018). Serum samples were tested with both, the iELISA-rRAP-1 and the iELISA-rMSA-1 tests for anti-*B. bigemina* antibodies and anti-*B. bovis* antibodies, respectively. The results obtained in the ELISA test were compared to those obtained in parallel testing with the IFAT method, currently considered as the gold standard test in Mexico (Alvarez et al. 2019).

RESULTS AND DISCUSSION

Descriptive Statistics

Table 1 presents descriptive statistics of the estimated diagnostic capacity for the *B. bigemina* iELISA test using rRAP-1 as antigen. The diagnostic sensitivity and specificity rates were 90% and 93.75%, respectively. While for the IFAT test the diagnostic sensitivity was 86.66% and the specificity was 95% (Table 2).

Table 1. Diagnostic sensitivity and specificity rates of the iELISA using *Babesia bigemina* rRAP-1 as antigen

Test	***B. bigemina* infection**			
iELISA		**Positive**	**Negative**	**Total Samples**
	Positive	27	5	32
	Negative	3	75	78
	Total Samples	30	80	110
Sensitivity (27/30) = 90%				
Specificity (75/80) = 93.75%				

Table 2. Diagnostic sensitivity and specificity rates of the IFAT using *Babesia bigemina* infected erythrocytes smears as antigen

Test	***B. bigemina* infection**			
IFAT		**Positive**	**Negative**	**Total Samples**
	Positive	26	4	30
	Negative	4	76	80
	Total Samples	30	80	110
Sensitivity (26/30) = 86.6%				
Specificity (76/80) = 95%				

Table 3. Diagnostic sensitivity and specificity rates of the iELISA using *Babesia bovis* rMSA-1 as antigen

Test	*B. bovis* infection			
iELISA		**Positive**	**Negative**	**Total Samples**
	Positive	24	1	25
	Negative	6	29	35
	Total Samples	30	30	60
Sensitivity (24/30) = 80%				
Specificity (29/30) = 97%				

Table 4. Diagnostic sensitivity and specificity rates of the IFAT using *Babesia bovis* infected erythrocytes smears as antigen

Test	*B. bovis* infection			
IFAT		**Positive**	**Negative**	**Total Samples**
	Positive	20	7	27
	Negative	10	23	33
	Total Samples	30	30	60
Sensitivity (20/30) = 66.7%				
Specificity (23/30) = 76.7%				

Table 3 depicts the descriptive statistics of the estimated diagnostic capacity for the *B. bovis* iELISA test using rMSA-1 as antigen. The diagnostic sensitivity and specificity rates were 80% and 97%, respectively. Whereas for the IFAT the sensitivity was 66.7% and the specificity of 76.7% for *B. bovis* antibodies, respectively (Table 4).

Concerning the analysis carried out with iELISA tests with recombinant antigens, *B. bovis* MSA-1 and *B. bigemina* RAP-1 performed on the 103 serum samples of the bovine herd at the "La Posta" Experimental Station, the following results were obtained: 23 serum samples (22.34%) resulted negative to both, the rMSA-1 and the rRAP-1 iELISA tests performed; 29 samples (28.15%) tested positive for iELISA-rMSA-1 (*B. bovis*); 14 samples (13.59%) tested positive in the iELISA-rRAP-1 (*B. bigemina*), while 37 samples (35.92%) were identified as positives in both tests (Figure 1).

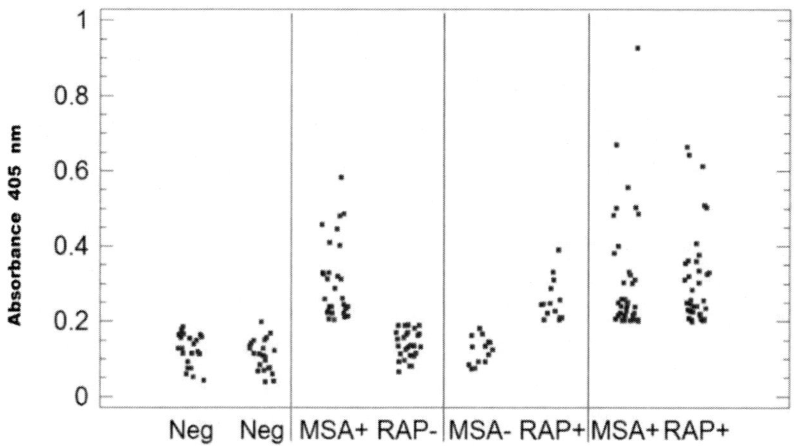

Figure 1. Results of the iELISA tests using *B. bovis* MSA-1 and *B. bigemina* RAP-1 recombinant proteins as antigens on bovine samples at the "La Posta" Experimental Station. Overall positivity to *Babesia* spp. is 77.66%, consisting of *B. bovis* MSA (28.15%); *B. bigemina* RAP (13.59%); *B. bovis* MSA + *B. bigemina* RAP (35.92%).

Table 5. iELISA and IFAT diagnostic tests concordance results between the iELISA RAP-1 and IFAT for *B. bigemina*

Test	iELISA RAP-1			
IFAT		**Positive**	**Negative**	**Total Samples**
	Positive	23	7	30
	Negative	7	73	80
	Total Samples	30	80	110
κ value = 0.68				
Concordance = Substantial				

Table 6. iELISA and IFAT diagnostic tests concordance results between the iELISA MSA-1 and IFAT for *B. bovis*

Test	iELISA MSA-1			
IFAT		**Positive**	**Negative**	**Total Samples**
	Positive	26	4	30
	Negative	4	26	30
	Total Samples	30	30	60
κ value = 0.73				
Concordance = Substantial				

Concordance (Kappa) Values Results

iELISA and IFAT diagnostic tests concordance results are reported in Tables 5 and 6. When determining the kappa concordance value between the iELISA RAP-1 and IFAT for *B. bigemina* antibodies, it was observed that the result was 0.68, which indicates that the agreement between the results obtained with the ELISA-rRAP-1 and the IFAT is substantial for the detection of anti-*B. bigemina* antibodies (Table 5). The estimated kappa value for the agreement between iELISA-rMSA-1 and IFAT for *B. bovis* was 0.70, indicating that it was substantial (Table 6).

A number of serological tests exist and practically all of them have been used in the diagnosis of bovine babesiosis (Bose et al. 1995; Maharana et al., 2016; Alvarez et al., 2019). The greater use of some of them is basically due to the apparent ease of obtaining the necessary reagents in the laboratory where these tests are performed (Bose, 1995, Alvarez et al., 2019). The most frequently used technique for the diagnosis of asymptomatic animals carrying *Babesia* is the Indirect Fluorescent Antibody Test (IFAT). Due to its relatively high sensitivity and the fact that *Babesia* hemoparasites have common antigens, cross-reaction issues become a problem with all *Babesia* species infecting cattle, requiring the establishment of a specific minimum reaction titration for each antigen-antibody system (Bose et al. 1995; Maharana et al., 2016; Alvarez et al., 2019).

In order to perform serosurveillance of bovine babesiosis, several immunological tests have been implemented in epidemiological studies. The most popular test used in the past is the indirect fluorescent antibody test (IFAT) considered as the gold test and characterized by its relatively high sensitivity and diagnostic specificity (≥ 90%) (Bautista-Garfias et al., 2012), The indirect immunoenzymatic assay (iELISA), like the IFAT, also allows the detection of circulating antibodies in the blood, among the main advantages of this technique are the ability to process a large number of samples during a working day, the objectivity of the results and the high sensitivity and diagnostic specificity (OIE, 2021). The determining factor or most important reagent to perform the serological assays is undoubtedly the antigen when it comes to detecting antibodies in serum samples (Bose et al., 1995; OIE 2021). Previous studies conducted in Mexico reported that the use of native crude antigens obtained from *in vitro* culture-derived *Babesia* parasites could increase the sensitivity and specificity of the ELISA test for the detection of anti-*B. bigemina* antibodies (Figueroa et al., 1992). In recent years, the problem associated with the lack of specificity of current serological assays has been solved by designing more sensitive tests using purified antigens (Bose, 1995; OIE, 2021). Thus, the study of recombinant proteins as antigens and their use in serological tests has been conducted in several laboratories around the world as a diagnostic alternative. A number of immunological diagnostic methods using recombinant proteins have been developed, which are mainly based on the identification of circulating antibodies produced by the host to babesial antigens (Boonchit, et al., 2004; Goff et al., 2003; 2008; Altangerel et al., 2009). Indirect ELISA and competition ELISA (C-ELISA) with recombinant merozoite proteins of *B. bovis* and *B. bigemina* have replaced the IFAT for the detection of antibodies against *Babesia* sp., due to the high processing efficiency and objectivity in the interpretation of results (OIE, 2021). However, no such assay is currently available in Mexico. Therefore, in the present study, research was conducted in order to assess and compare the IFAT and iELISA with species-specific *B. bovis* and *B. bigemina* recombinant proteins as antigens, work in which the agreement between the IFAT and iELISA was determined in order to obtain a better serodiagnosis of bovine babesiosis in Mexico.

In this study, the estimated specificity of the iELISA tests using *B. bigemina* RAP-1 and *B. bovis* MSA-1 recombinant antigens was slightly higher than the sensitivity because the cut-off points with which each of the results were classified was determined based on the Absorbance at OD 405 nm results of the truly negative samples. In addition, the results of this study show lower sensitivity for the IFAT test, essentially due to the subjectivity that usually occurs when performing microscopic analysis of fluorescent cells (Alvarez et al., 2019). Once the concordance value was calculated, it was determined that the 0.68 kappa value, indicates that the match between the results obtained with the *B. bigemina* iELISA-rRAP-1 and IFAT tests is substantial for the detection of anti - *B. bigemina* antibodies. The estimated kappa value for the concordance between iELISA-rMSA-1 and IFAT for *B. bovis* was 0.73, indicating that, like the previous assay, it was also substantial (Landis & Koch, 1977). The lack of concordance in the iELISA and IFAT tests for *B. bigemina* was essentially based on the results obtained with 7 positive serum samples in the iELISA test which were negatives in the IFAT test, as well as on the results obtained with 7 different positive serum samples in the IFAT test, that were negative in the iELISA test. It is likely that this result is rather associated with an antibody crossover reaction with antigenic molecules present in *B. bigemina* like those present in *B. bovis* since the 7 discordant serum samples reacted with *B. bovis*-infected erythrocytes in the parallel IFAT test performed. Notwithstanding these results, the iELISA instrumented with rRAP-1 and rMSA-1 as antigens have an adequate diagnostic sensitivity and specificity rate which, once validated with field samples and applied in a broader epidemiological study, would allow the detection of cattle naturally exposed to infection with *B. bigemina* and/or *B. bovis*.

To complement the present study and with the use of the recombinant antigens in the iELISA, the determination of seroprevalence was performed in an area where the animals are in constant exposure to the causative agent of babesiosis. In this sense, the results of the study carried out in the bovine herd located in Veracruz, Mexico, have allowed us to determine the usefulness of the iELISA tests evaluated. Thus, the aggregate prevalence rate of animals with antibodies to *Babesia* spp determined in the bovine herd using both iELISA tests using the 2 different recombinant antigens was estimated at 76.6%, slightly higher than that obtained with the IFAT (71%), considered as the standard test for estimation of seroprevalence to bovine babesiosis in Mexico (Alvarez et al., 2017, 2019; Esteve-Gasent et al., 2020). The results obtained in the analysis of the samples of the present work in terms of seroprevalence indicate that the herd of the "La Posta" Experimental Field Station is located in an endemic area of bovine babesiosis. The seroprevalences obtained through the use of the iELISA test, as well as the IFAT, are considered high and coincide with those reported previously (Lira et al., 2015), in regions considered of high endemicity in Mexico (Veracruz and Tamaulipas), where seroprevalence rates of 90.2% in the case of *B. bovis* and 83.7% for *B. bigemina* were obtained in a cattle herd located in Tamaulipas, meanwhile, the herd located in the state of Veracruz showed a seroprevalence of 73.1% for *B. bigemina* and 64.5% for *B. bovis*. The latter is considered as a relatively low seroprevalence rate since, in previous studies, a prevalence of up to 80% was determined for *Babesia* spp (Álvarez et al. 2007). The low prevalence of circulating antibodies in the samples of the state of Veracruz could be explained due to the application of a comprehensive control program of the vector, the tick *Rhipicephalus (Boophilus) microplus*.

The seroprevalences rates for *B. bovis* species (64.5% and 90.2%) and *B. bigemina* (73.1% and 83.7%) in animals raised in endemic areas such as the States of Veracruz and Tamaulipas (Lira et al., 2015) as well as those shown in the present study, show high seroprevalence rates, as compared to the seroprevalence rates found in areas considered of low endemicity, such as that reported in the state of Morelos, Mexico (Rojas et al., 2004), in which seroprevalences of 58% and 60% for *B. bovis* and *B. bigemina*, respectively, were determined by using the IFAT.

The determination of enzootically stable or unstable areas for bovine babesiosis through the measurement of total IgG antibodies specific to *Babesia* sp in animals between 3 and 9 months of age, has been used as an epidemiological indicator (Guglielmone, 1995; Regassa et al., 2003). Thus, an area presents enzootic stability for bovine babesiosis when 75% or more of cattle between 3 and 9 months of age are seroreactive to *Babesia* (Guglielmone, 1995; Regassa et al., 2003).

CONCLUSION

The instrumented iELISA tests have an excellent diagnostic specificity rate and acceptable sensitivity that allow the detection of cattle previously exposed to *Babesia bovis*, *B. bigemina,* or both. Nonetheless, it still requires extensive validation with field samples from different regions in Mexico to be suggested as a validated assay to be included in a disease prevention and control program that facilitates the estimation of the daily probability rate of bovine babesiosis infection. The estimation of this infection rate based on the values obtained will allow the establishment of appropriate criteria for the intervention and use of a program for the prevention and control of bovine babesiosis in Mexico.

Acknowledgments

This study was partially funded by CONACYT, Mexico, Problemas Nacionales 2015, Project No. 1336; SEP-CONACYT CB2017-2018 Project No. A1-S-43508; and INIFAP, Project No. 1672534936.

References

Altangerel, K., Alhassan, A., Iseki, H., Sivakumar, T., Boldbaatar, D., Yokoyama, N. and Igarashi, I. (2009). Evaluation of *Babesia bigemina* 200 kDa recombinant antigen in enzyme-linked immunosorbent assay. *Parasitology Research*, *105*(1), 249-54.

Álvarez-Martinez, J. A. and Cantó-Alarcón, G. J. (1985). Epidemiología de la babesiasis. En: *Sociedad Mexicana de Parasitología (ed.). Parasitología, Vol. I Conmemorativo 25 Aniversario.* México, D.F. pp. 54-72. [Epidemiology of babesiasis. In: *Mexican Society for Parasitology (ed.). Parasitology,* Vol. *I Commemorative 25th Anniversary*. Mexico D.F. pp. 54-72].

Alvarez, J. A., Ramos, J. A., Rojas, E. E., Mosqueda, J. J., Vega, C. A., Olvera, A. M., Figueroa, J. V. and Cantó, G. J. (2004). Field challenge of cattle vaccinated with a combined *Babesia bovis* and

Babesia bigemina frozen immunogen. *Annals of the New York Academy of Sciences*, *1026*, 277-283.

Álvarez, J. A., Alpírez, M. F., Rojas, M. C. and Figueroa, J. V. (2007). Probabilidad diaria de infección para *Babesia* spp mediante PCR anidado. En: INIFAP, UV, CP, UACH, ITUG, ITBOCA, UNAM (eds): *Avances en la Investigación Agrícola, Pecuaria, Forestal y Acuícola en el Trópico Mexicano, Veracruz, México*, *4*, 397-402. [Daily probability of infection for *Babesia* spp by nested PCR. In: INIFAP, UV, CP, UACH, ITUG, ITBOCA, UNAM (eds): *Advances in Agricultural, Livestock, Forestry and Aquaculture Research in the Mexican Tropics, Veracruz, Mexico*, *4*, 397-402].

Alvarez, J. A., Lopez, U., Rojas, C., Borgonio, V. M., Sanchez, V., Castañeda, R., Vargas, P. and Figueroa, J. V. (2010). Immunization of *Bos taurus* steers with *Babesia bovis* recombinant antigens MSA-1, MSA-2c and 12D3. *Transboundary and Emerging Diseases*, *57*, 87-90.

Alvarez, J. A., Rojas, C. and Figueroa, J. V. (2019). Diagnostic tools for the identification of *Babesia* sp. in persistently infected cattle. *Pathogens*, *8*, 143.

Alvarez, J. A., Rojas, C. and Figueroa, J. V. (2020). An overview of current knowledge on *in vitro Babesia* cultivation for production of live attenuated vaccines for bovine babesiosis in Mexico. *Frontiers in Veterinary Science*, *7*, 364.

Bautista-Garfias, C. R., Castañeda-Arriola, R., Álvarez-Martínez, J. A., Rojas-Martánez, C., Figueroa-Millán, J. V. and Rodríguez-Lozano, A. (2012). La vacunación simultánea de bovinos con *Lactobacillus casei* y la vacuna bivalente contra babesiosis bovina genera una mejor protección contra *Babesia bovis* y *B. bigemina* transmitidas por garrapatas en condiciones extremas de campo. *Revista Veterinaria México*, *43*(3), 198-200. [The simultaneous vaccination of bovines with *Lactobacillus casei* and the bivalent vaccine against bovine babesiosis induces a better protection against *Babesia bovis* and *B. bigemina* transmitted by ticks in extreme field conditions. *Veterinary Mexico Magazine*, *43*(3), 198-200].

Bock, R., Jackson, L., de Vos, A. and Jorgensen, W. (2004). Babesiosis of cattle. *Parasitology*, *129*, 247-269.

Boonchit, S., Xuan, X., Yokoyama, N., Goff, W. L., Waghela, S. D., Wagner, G. and Igarashi, I. (2004). Improved enzyme-linked immunosorbent assay using C-terminal truncated recombinant antigens of *Babesia bovis* rhoptry-associated protein-1 for detection of specific antibodies. *Journal of Clinical Microbiology*, *42*, 1601-1604.

Borgonio, V., Mosqueda, J., Genis, A. D., Falcon, A., Alvarez, J. A., Camacho, M. and Figueroa, J. V. (2008). *msa-1* and *msa-2c* gene analysis and common epitopes assessment in Mexican *Babesia bovis* isolates. *Annals of the New York Academy of Sciences*, *1149*, 145-148.

Bose, R., Jorgensen, W. K., Dalgliesh, R. J., Friedhoff, K. T. and de Vos, A. J. (1995). Current state and future trends in the diagnosis of babesiosis. *Veterinary Parasitology*, *57*, 61-74.

Esteve-Gasent, M. D., Rodríguez-Vivas, R. I., Medina, R. F., Ellis, D., Schwartz, A., Cortés- Garcia, B., Hunt, C., Tietjen, M., Bonilla, D., Thomas, D., Logan, L. L., Hasel, H., Alvarez-Martínez, J. A., Hernández-Escareño, J. J., Mosqueda-Gualito, J., Alonso-Díaz, M. A., Rosario-Cruz, R., Soberanes-Céspedes, N., Merino-Charrez, O., Howard, T., Chávez-Niño, V. M. and Pérez de León, A. A. (2020). Research on integrated management for cattle fever ticks and bovine babesiosis in the United States and Mexico: Current status and opportunities for binational coordination. *Pathogens*, *9*(11), 871.

Figueroa, J. V., Buening, G. M., Kinden, D. A. and Green, T. J. (1990). Identification of common surface antigens among *Babesia bigemina* isolates by using monoclonal antibodies. *Parasitology*, *100*, 161-175.

Figueroa, J. V., Alvarez, J. A., Buening, G. M., Canto, G. J, Hernandez, R., Monroy, M., Ramos, J. A. and Vega, C. A. (1992). Antibody response to *Babesia bigemina* infection in calves measured by ELISA and immunoblotting techniques. *Revista Latinoamericana de Microbiologia*, *34*, 267-273.

Figueroa, J. V., Chieves, L. P., Johnson, G. S. and Buening, G. M. (1993). Multiplex polymerase chain reaction-based assay for the detection of *Babesia bigemina*, *Babesia bovis* and *Anaplasma marginale* DNA in bovine blood. *Veterinary Parasitology*, *50*, 69-81.

Figueroa, J. V. and Buening, G. M. (1995). Nucleic acid probes as a diagnostic method for tick-borne hemoparasites of veterinary importance. *Veterinary Parasitology*, *57*, 75-92.

Figueroa, J. V., Kinden, D. A. and Buening, G. M. (1998). Use of monoclonal antibodies for the identification of a common surface antigen of *Babesia bovis*. *Annals of the New York Academy of Sciences*, *849*, 433-437.

Florin-Christensen, M., Suarez, C. E., Hines, S., Palmer, G. H. and McElwain, T. F. (2002). The msa-2 locus of *Babesia bovis* contains four tandemly arranged, co-expressed genes encoding genetically related but immunologically distinct surface proteins. *Infection & Immunity*, *70*, 3566–3575.

Goff, W. L., Davis, W. C., Palmer, G. H., McElwain, T. F., Johnson, W. C., Bailey, J. F. and McGuire, T. C. (1988). Identification of *Babesia bovis* merozoite surface antigens by using immune bovine serum and monoclonal antibodies. *Infection & Immunity*, *56*, 2363-2368.

Goff, W. L., McElwain, T. F., Suares, C. E., Johnson, W. C., Brown, W., Norimine, J. and Knowles, D. P. (2003). Competitive enzyme-linked immunosorbent assay based on a rhoptry-associated protein 1 epitope specifically identifies *Babesia bovis*-infected cattle. *Clinical and Diagnostic Laboratory Immunology*, *10*, 38-43.

Goff, W. L., Johnson, W. C., Molloy, J. B., Jorgensen, W. K., Waldron, S. J., Figueroa, J. V., Matthee, O., Adams, D. S., McGuire, T. C., Pino, I., Mosqueda, J., Palmer, G. H., Suarez, C. E., Knowles, D. P. and McElwain, T. F. (2008). Validation of a competitive enzyme-linked immunosorbent assay for detection of *Babesia bigemina* antibodies in cattle. *Clinical and Vaccine Immunology*, *15*, 1316-21.

Guglielmone, A. A. (1995). Epidemiology of babesiosis and anaplasmosis in South and Central America. *Veterinary Parasitology*, *57*(1-3), 109-119.

Hines, S., McElwain, T., F., Buening, G. and Palmer, G. H. (1989). Molecular characterization of *Babesia bovis* merozoite surface proteins bearing epitopes immunodominant in protected cattle. *Molecular & Biochemical Parasitology*, *37*, 1-9.

Landis, J. R. and Koch, G. G. (1977). The measurement of observer agreement for categorical data. *Biometrics*, *33*, 159-174.

Lira, J. J., Vargas, U. P., Cantú, C. A., Castañeda, A. R. O., Álvarez, J. A., Rojas, M. C., Bautista, C. R. and Figueroa, J. V. (2015). Prevalencia de babesiosis bovina en dos explotaciones del país utilizando prueba serológica con antígenos recombinantes. In: Martìnez-Herrera, J. (ed), *Memorias de XXVII Reunión Científica Tecnológica, Forestal y Agropecuaria, Tabasco, Mexico*. pp. 217-220. [Prevalence of bovine babesiosis in two farms in Mexico using serological testing with recombinant antigens. *Proceedings XXVII Technological, Forestry and Agricultural Scientific Meeting, Tabasco, Mexico*. pp. 217-220].

Martinez-García, G., Santamaria-Espinosa, R. M., Lira-Amaya, J. J. and Figueroa, J. V. (2021). Challenges in tick-borne pathogen detection: the case for *Babesia* spp. identification in the tick vector. *Pathogens*, *10*, 92.

Maharana, B. R., Tewari, A. K., Saravanan, B. C. and Sudhakar, N. R. (2016). Important hemoprotozoan diseases of livestock: Challenges in current diagnostics and therapeutics: An update. *Veterinary World*, *9*, 487-495.

McCosker, P. J. (1981). The global importance of babesiosis. In: Ristic, M. and Kreier, J. P., (eds.) *Babesiosis.* Academic Press, New York, pp. 1-24.

McElwain, T. F., Perryman, L. E., Musoke, A. J. and McGuire, T. C. (1991). Molecular characterization and immunogenicity of neutralization sensitive *Babesia bigemina* merozoite surface proteins. *Molecular and Biochemical Parasitology*, *47*, 213-222.

OIE. (2021). Bovine babesiosis. In *OIE Terrestrial Manual*, Chapter 3.4.2. Version adopted in May 2021. pp 1-16. Consulted in May 2021.

Regassa, A., Penzhorn, B. and Bryson, N. (2003). Attainment of endemic stability to *Babesia bigemina* in cattle on a South African ranch where non-intensive tick control was applied. *Veterinary Parasitology*, *116*, 267-274.

Rojas, R. E., Domínguez, P., García, M., Cruz, V. C., Figueroa, J. V. and Ramos, A. J. (2004). Prevalencia e incidencia de *Babesia bovis* y *Babesia bigemina* en un hato bovino en Axochiapan, Morelos. *Avances en Investigación Agropecuaria*, *8*, 1-8. [Prevalence and incidence of *Babesia bovis* and *Babesia bigemina* in a cattle herd in Axochiapan, Morelos. *Advances in Agricultural Research*, *8*, 1-8].

Rojas-Martínez, C., Rodríguez-Vivas, R. I., Figueroa-Millán, J. V., Bautista-Garfias, C. R., Castañeda-Arriola, R. O., Lira-Amaya, J. J., Vargas-Urióstegui, P., Ojeda-Carrasco, J. J. and Alvarez-Martínez, J. A. (2018). Bovine babesiosis: Cattle protected in the field with a frozen vaccine containing *Babesia bovis* and *Babesia bigemina* cultured *in vitro* with a serum-free medium. *Parasitology International*, *67*, 190–195.

Santamaria, R. M., Lira, J. J., Vargas, P., Alvarez, J. A., Rojas, C. and Figueroa, J. V. (2020). Validation of an indirect ELISA using recombinant proteins as antigen to identify animals exposed to *Babesia bigemina*. *Transboundary and Emerging Diseases*, *67*(S2), 201-207.

Simoes, C. D. A., Chirinos, R. A. R., Martinez, C. N. S., Castejon, O. and Avila, J. P. (1995). Prevalence of bovine babesiosis in Sector Cuatro (Playa Bonita). Mara Country. *Revista Científica FVC-LUZ*, *5*(1), 5-10. [Prevalence of bovine babesiosis in Sector Four (Playa Bonita). Mara Country. *Scientific Journal FVC-LUZ*, *5*(1), 5-10].

In: ELISA
Editor: Jose Ma. M. Angeles
ISBN: 978-1-68507-500-2

Chapter 4

ELISA: A SERO-SURVEILLANCE TOOL IN MALARIA ELIMINATION

Pilarita T. Rivera*
Department of Parasitology, College of Public Health,
University of the Philippines Manila, Manila, Philippines

ABSTRACT

The pursuit of malaria elimination has set malaria surveillance as a core intervention in all countries endemic and malaria-free but at risk of malaria re-introduction. There is thus a need for malaria surveillance tools that can be utilized by the malaria programs of these countries. This review aimed to describe malaria enzyme-linked immunosorbent assay (ELISA) as a sero-surveillance tool in malaria elimination.

Scrological tests have been used in seroepidemiological surveys to assess malaria transmission intensity (MTI), and measure reduction or confirm interruption of malaria transmission. ELISA has been the most widely established and used method because it is efficient, high-throughput, and low cost. Several recombinant antigens of the malaria parasite, *Plasmodium* have been produced and used in the ELISA assay.

* Corresponding Author's E-mail: ptrivera@up.edu.ph.

As a malaria sero-surveillance assay, ELISA using recombinant antigens has been done to measure malaria seroprevalence, seroconversion, and sero-reversion as indicators of MTI, malaria transmission hotspots, and malaria transmission dynamics. But while the utility of ELISA as a sero-surveillance assay has already been demonstrated, a point of care (POC) serologic test can make malaria surveillance truly helpful and sustainable.

Keywords: ELISA, malaria, surveillance

INTRODUCTION

Serology or the detection and measurement of anti-malarial antibodies has been used as an epidemiological tool. Sero-epidemiological studies have previously been done to assess malaria transmission intensity, measure reductions in transmission, and confirm interruption of transmission. However, variation in the source of antigen and the subjectivity of the detection methods became the major challenges to their utility. But, the subsequent availability of specific recombinant malarial antigens and the development of standardized, sensitive enzyme-linked immunosorbent assays (ELISA) are once again proving the value of seroepidemiology both for assessment of malaria transmission, for measurement of changes in prevalence following the implementation of control programs and confirmation of malaria elimination (Cook et al., 2010; Tanner et al., 2015).

Serologic tests have been used to detect variations in malaria transmission, and thus, monitor trends in malaria endemicity, evaluate the effectiveness of malaria interventions and control programs (Drakely et al., 2005), and measure malaria transmission intensity (Corran et al., 2007). Cook and co-workers also pointed out that when the transmission is low, the conventional measures of malaria transmission intensity (MTI) such as the entomological inoculation rate (EIR) and parasite rates (PR) lack sensitivity.

This is because of the low numbers of positive mosquitoes and humans at such a level of transmission. In addition, the low frequency of positive samples is a difficult logistical challenge. An alternative measure for MTI is to compute the prevalence of anti-malaria antibodies in the local population. Serological markers of transmission show greater sensitivity in low transmission areas, as seroprevalence reflects cumulative exposure and specific antibody responses are of longer duration. Thus, seroprevalence is less affected by seasonality in contrast to entomological and parasitological measures that are very much affected by seasonality. Blood sample collection and analysis can be readily scaled up, and analysis is relatively simple, making this approach adaptable for resource-poor settings. Serological measures are robust and high throughput and importantly, remain sensitive in very low transmission settings (Cook et al., 2010).

It has also been noted that in areas of low malaria transmission intensity, serological indicators could be used to infer risk factors for exposure and to assess changing transmission dynamics across the season and, also over time and space. For long-term monitoring of exposure as transmission levels drop further, serology may be a valuable tool, as a detailed examination of age-specific seroprevalence data and seroconversion rates (SCRs) can be used to monitor changes in transmission, and to detect transmission hot-spots (Baum et al., 2015). Measurement of SCR's from individual-level longitudinal data of antibody titers, has made possible the identification of individual seroconversion, which can be useful in assessing interruption of malaria transmission (Bretscher et al., 2013). In areas where malaria elimination protocols have been implemented, serological measures, in combination with molecular parasitological measures, can help to confirm areas where the infection is no longer present (Cook et al., 2010). The absence of antibodies against *Plasmodium* has been used to show the success of elimination programs in Mauritius, Greece, and in Vanuatu (Baum et al., 2015).

Applications of Serological Surveillance in Malaria Control and Elimination

Sero-surveillance is the use of *Plasmodium* species-specific antibodies as biomarkers for monitoring exposure, transmission, and immunity, and has significant potential to enhance the effectiveness of malaria control and elimination programs. Antibodies are a more sensitive marker of population-level malaria exposure in low-transmission settings and reflect exposure over a period of time, which is useful in areas with highly seasonal transmission. Although this approach was used historically as part of malaria control programs, its use declined in part because of the lack of standardized antigens and methodology. Recently, with the availability of recombinant proteins and improved technology for high-throughput screening, there is renewed interest in sero-surveillance for malaria (Elliott et al., 2014).

Several studies have explored the measurement of serum antibodies to recombinant malaria proteins by enzyme-linked immunosorbent assay (ELISA), as a measure of malaria exposure, and a number of potential applications for sero-surveillance have been identified. For instance, seroprevalence provides a general indication of the level of endemicity, with a peak in seroprevalence at an earlier age in areas with high transmission intensity. Modeling of age-specific seroprevalence data may be used to estimate the seroconversion rate or the rate at which seronegative individuals become seropositive. This seroconversion has been shown to be associated with measures of transmission intensity, such as malaria incidence and EIR (Stewart et al., 2009; Bretscher et al., 2013). Also, by determining seroprevalence, trends in malaria transmission over time, and age-specific antibody prevalence rates may be identified and used to estimate the time at which major changes in transmission occurred.

Sero-surveillance is a more sensitive tool than parasite detection by microscopy of blood smears or rapid diagnostic tests in evaluating the impact of control interventions on transmission, particularly when MTI is already low. Sero-surveillance has also proven to be a valuable method for identifying transmission hotspots and population groups at higher risk of infection. As transmission decreases, it becomes more heterogeneous, concentrated in defined geographic areas ("hotspots") and high-risk subpopulations ("hot-pops"). At the same time, the proportion of individuals with sub-microscopic infection increases, consequently contributing substantially to ongoing transmission and posing a challenge to the identification and elimination of persisting malaria reservoirs (Bousema et al., 2013).

Sero-surveillance could be used to verify that malaria has been eliminated since the absence of antibodies in the youngest age groups indicates that malaria transmission has been interrupted. Post-elimination, there remains a need to screen mobile populations at risk of transmission. Serology is unlikely to be useful for diagnosing actively infected individuals because antibodies take days to develop and persist after an infection has resolved, but serology might identify high-risk groups suitable for active case detection. Sero-surveillance might also identify populations at risk of outbreaks as immunity wanes in the face of declining malaria transmission, but this will require further research to identify robust correlates of anti-malarial immunity. Sero-surveillance is likely to be used mainly for monitoring the transmission of, and exposure to *Plasmodium falciparum* and *Plasmodium vivax* and may prove helpful for identifying individuals with *P. vivax* hypnozoites (which are responsible for *P. vivax* relapses), although a suitable serological test is required (Elliott et al., 2014).

Development of Serological Assays for Malaria Surveillance

Elliot et al., proposed a model for the development of sero-surveillance assays and tests, that outlined the key points necessary for their utility: the purpose of doing the sero-surveillance assay, its technical performance, its development and optimization, its validation and application in the reference laboratory, and the field. In terms of assay development, the selection of antigens to be used in the assay is a primary consideration.

The potential of antigens for malaria sero-surveillance assays depends on their immunogenicity, antibody longevity, polymorphism, and cross-reactivity between *Plasmodium* species. For instance, highly immunogenic antigens induce antibodies that are acquired early in life or after limited malaria exposure and will be well used in assays for monitoring changes in transmission in low-endemicity settings. While antigens that induce more stable antibody responses are ideal for assays measuring cumulative exposure and assessing changes in transmission over extended periods of time, those of shorter duration will be more useful for identifying recent exposure. The target population and application of the assay are also considered in the choice of antigen. While highly immunogenic antigens will be preferable for a target population of young children to confirm malaria elimination, antigens with short-lived antibody responses will be needed in assays to identify subpopulations of adults with recent occupational exposure. Also, antigens that are species-specific or those that show substantial sequence diversity between species and limited antibody cross-reactivity are ideal in ensuring high assay sensitivity and specificity. Finally, antigens that are given as vaccines cannot be utilized in sero-surveillance because of the difficulty in differentiating vaccine-induced from naturally acquired antibodies (Corran et al., 2007; Elliott et al., 2014).

The review of Elliott et al., listed 3 antigens of both *P. falciparum* and *P. vivax* namely, merozoite surface protein 1 (MSP-1), apical membrane antigen (AMA-1), and circumsporozoite protein (CSP), and whose utility has been demonstrated in sero-surveillance. However, CSP and AMA-1 being candidate malaria vaccines may not eventually, be the antigens of choice for sero-surveillance assays. The list also included *P. falciparum* specific antigens- erythrocyte binding antigen (EBA-175), merozoite surface protein 2 (MSP-2), *P. falciparum* reticulocyte-binding homolog 2 (PfRH-2) and variant surface antigen (VAR2CSA), and *P.*

vivax specific antigens- Duffy binding protein (DBP) and reticulocyte-binding protein 1 (RBP-1). Baum et al., detected antibody responses to well-known antigens, such as several *P. falciparum* merozoite surface proteins (MSP), AMA-1, liver-stage antigen 1 (LSA-1), serine repeat antigens (SERA) 3, 4 and 5; erythrocyte membrane protein 1 (PfEMP-1); and *P. vivax* CSP (PvCSP), early transcribed membrane proteins (ETRAMP), multiple MSP and SERA proteins (Baum et al., 2015).

A systematic review on the identification of serologic markers of vivax malaria exposure and protective immunity showed that IgG responses to PvCSP, PvMSP-1_{19}, Pv MSP-9_{NT}, and PvAMA1 were associated with increased odds of *P. vivax* infection in geographically diverse populations. Other antigens (PvMSP-3α, PvMSP-9RIRII, Pv DBP, and PvRBP-1) were also shown to be markers of *P. vivax* infection, but only in single populations. Sero-surveillance using PvCSP in Korea and PvMSP-1_{19} and PvAMA1 in Vanuatu, Cambodia, and Somalia had been employed to successfully map *P. vivax* transmission (Cutts et al., 2014). An earlier systematic review and meta-analysis on antibody responses to blood-stage antigens of *P. falciparum* showed IgG antibody responses to MSP-1 (including processing fragments and defined blocks), MSP-2, and MSP-3, AMA-1, EBA-175, and GLURP in the ELISA (Fowkes et al., 2010).

Considering antigens of other *Plasmodium* species, Herman et al., noted that for *Plasmodium knowlesi*, PK66 (Pk AMA1) and Pk SPATR (secreted protein with altered thrombospondin repeat) share 86% and 85% amino acid identity respectively with *P. vivax*, potentially making it difficult to distinguish between the two species where infections are co-endemic. Testing serum samples for immunoreactivities to other *P. knowlesi* recombinant proteins- cysteine protease (serine repeat-like antigen) SERA3ag1, SERA3ag2; sporozoite surface protein 2, putative, thrombospondin-related anonymous protein (TRAP); and truncated cysteine protease TSERA2ag1, the results showed SERA3ag2 as the most promising *P. knowlesi* antigen in sero-surveillance assays (Herman et al., 2018).

More recently, Muller Sienerth et al., expanded this list of *Plasmodium* antigens to include those proteins of the other species- *Plasmodium malariae*, *Plasmodium ovale,* and *P. knowlesi*. The immunoreactivities of antibodies IgG, IgM, and IgA from serum samples of individuals diagnosed to be infected with any of these 5 *Plasmodium* species were determined by ELISA. The results showed that immunoreactivities to *P. ovale* merozoite surface protein (MSP10) and 6-cysteine protein, putative (P41), *P. malariae* MSP10, and *P. knowlesi* MSP10, P12, P38 were able to predict prior exposure to these species of *Plasmodium*. As a result, this expanded panel of antigens may contribute to an assay that will be a useful tool for establishing prior exposure to different species of *Plasmodium* parasites (Muller Sienerth et al., 2020).

With regards to the serology platforms or methodologies for sero-surveillance assays, Elliott et al., listed ELISA, suspension bead arrays, and protein microarrays. However, ELISA has been the most widely established and used method, because it is efficient, high-throughput, and low cost. Although, it is laboratory-based, and semi-automated, limited technical expertise is required. However, its major drawback is limited testing to only a single antigen at a time. In contrast, bead arrays with antigens coated on fluorescent beads, are able to detect antibody responses to multiple antigens simultaneously, while microarrays detect antibodies to an extensive array of antigens. The problem is the feasibility of these methodologies outside of the laboratory setting (Elliott et al., 2014).

The most widely used ELISA uses recombinant antigens coated on plates at the concentration of 0.5 mg/mL buffer and incubated at 4°C overnight. The plates are washed in PBST and blocked with 1% (w/v) skimmed milk for 3 hours. After the washing steps, the serum samples are added in duplicate at a final dilution of 1:1,000 and incubated overnight at 4°C. The plates are then washed and 50μl of HRP-conjugated rabbit anti-human IgG is added into each well and incubated for 3 hours. After the washing steps, the substrate solution is added, and the reaction is allowed to develop for 15– 20 min before the addition of the stopping solution (2 M H_2SO_4). The optical density is read using an ELISA reader at 450 nm (Corran et al., 2007).

A previously described research-based (RB) ELISA protocol was able to capture the antibody responses in a single assay, using a pool of five *P. falciparum* antigens: apical membrane antigen 1 (AMA-1), the 19 kDa fragment of merozoite surface protein 1 (MSP-1_{19}), the full-length Dd2 allele of MSP-2 (MSP-2 Dd2), the full-length CH150/9 allele of MSP-2 (MSP-2 CH150/9) and glutamate-rich protein R2 (GLURP-R2). The coating concentration for all antigens was 0.5μg/ml, except for GLURP-R2 at 0.1μg/ml. To increase throughput and mimic commercial assays, samples were run in single wells. Samples were tested at a final concentration of 1:1,000 OD measures and read at a wavelength of 450 nm, corrected for blank responses (ODcorr), and normalized using the standard curve (van den Hoogen et al., 2020).

A comparison of the technical performance of 5 commercial malaria ELISA kits (Trinity Biotech, newbio, DiaPro, Cellabs, and NovaTec) with the RB ELISA, in testing as true positives the serum samples from a malaria hyperendemic region in the Gambia that were collected in the 1990s, revealed high sensitivity (90–95%) except for the DiaPro kit (86%). Specificity using samples from either malaria unexposed individuals or those infected by *Toxoplasma* was also high (>96%) except for the Cellabs kit (81 and 84%, respectively). With regards ease of use, the DiaPro kit needed the highest volume of serum (150 μl compared to ≤50 μl for the other kits) and was considered as the least user-friendly. Costs per sample were highest for the DiaPro, Cellabs, and NovaTec kits (> USD 2 compared to < USD 2 for the Trinity Biotech and newbio kits) On the other hand, the results of the assessment of the epidemiological performance of Trinity Biotech and NovaTec against the RB ELISA, in the characterization of malaria transmission at low endemicity and pre-elimination settings, showed that Trinity Biotech seroconversion curve profiles agreed with the decrease in passively detected malaria cases in Bataan, the Philippines in 1995. Moreover, the seroprevalence in those born since this decrease in transmission was low (1.3%). The results for Cape Verde, showed the low, constant seroconversion rate (SCR) as detected by the Trinity Biotech kit, and provided proof of the low levels of malaria transmission over the past decades. This kit also identified a higher exposure to malaria in adult men (i.e., >20 years old compared to those younger than 20 years old), and that matched previously described data in epidemiological studies in Cape Verde. The RB ELISA produced similar results but the NOVATec kit did not reflect malaria transmission patterns either in the Philippines and Cape Verde, and therefore it probably has limited utility for epidemiological characterization of transmission (van den Hoogen et al. 2020).

ELISA as a Malaria Sero-Surveillance Assay

ELISA detecting antibodies to the CSP had been used to determine MTI and cumulative (age-related) exposure to infection in Brazil and Sri Lanka (Corran et al., 2007). Meanwhile, anti-MSP-1_{19} and anti-AMA-1 human IgG antibodies as measured by ELISA were used to determine the age-specific seroprevalence and seroconversion rates for rapid assessment of MTI in 2 study areas in Tanzania. The SCR results showed that malaria transmission varied across the study areas, and there was a marked reduction of transmission in one study area where SCR was 5 to 10 fold lower among those born since 1998 compared to those born prior to this year. This study also showed that a health center-based serological survey is a rapid, cheap, and easy approach to generate data on MTI needed in the implementation and monitoring of malaria control programs (Stewart et al., 2009).

Using ELISA to estimate recent and long-term malaria transmission, antibodies to *P. falciparum* CSP ($NANP_5$ repeat peptide), and schizont extract crude antigen (SE) were detected in serum samples from individuals living in highly seasonal falciparum malaria transmission areas in Kenya. The results showed that CSP and SE specific IgG antibodies are acquired slowly with only <40% of the study participants as positive by 20 years of age. Moreover, the sero-reversion (the change from seropositive to seronegative) results showed that sero-reversion of CSP specific antibodies, slowly and linearly decrease with age, while SE specific antibodies showed high sero-reversion (>80%) in all ages. In addition, results showed that the CSP and SE antibody half-life was low in young children and did not increase with age, remaining at <2 years in adults aged >50 years. It could be concluded that depending on the antigens used, this sero-surveillance ELISA might estimate recent and cumulative *P. falciparum* exposure in a population, and even their future risk of clinical malaria (Ondigo et al., 2014).

P. falciparum glutamate-rich protein (GLURP) and *P. vivax* MSP-1_{19} ELISA were utilized to estimate MTI in areas with seasonal and residual malaria transmission along the Cambodia and Thailand border. The results showed that there was species-specific seroprevalence depending on the region and season (*P. falciparum* on the east, during the rainy season; *P. vivax* on the west, during the dry season). There was a linear increase in the proportion of study subjects seroconverting from less than 10% in the under-five years of age and up to 30% in the adults aged 25 to 40. These results may indicate the usefulness of serological indicators to infer risk factors for malaria exposure and to assess changing transmission dynamics across the season as well as overtime and space (Cook et al., 2012). A cross-sectional survey in the Peruvian Amazon included ELISA to characterize the current malaria transmission patterns and identify hotspots as well as recent changes in specie specific MTI. The seroprevalence to Pv MSP-1_{19} and AMA-1, and Pf GLURP and AMA-1 were 33.6% and 22% respectively. The spatial analysis identified well-defined clusters of *P. falciparum* seropositive individuals, while it detected only a very small cluster of *P. vivax* exposure (Rosas-Aguirre et al., 2015). In Korea, his tagged recombinant proteins coding the antigenic domains of *P. vivax* CSP-1, MSP-1, AMA-1, SERA, and EXP-1 were used in the ELISA and tested serum samples collected from study subjects in the Demilitarized Zone, a malaria-endemic area. The results showed a 7.2% seroprevalence that may indicate asymptomatic infection considering the prolonged incubation and dormant stages of *P. vivax* (Kim et al., 2016).

In Java Island, Indonesia, antibody responses against PfMSP-1_{19} and PfAMA-1 were used to assess differences in seroprevalence between a hypo-endemic and meso-endemic area, as well as seasonal changes in malaria exposure. The data showed that antibody responses, either as prevalence, SCR, or magnitude of antibody response were able to differentiate between areas of different transmission intensity, and between transmission seasons in each area (Supargiyono et al., 2013). While in another malaria-endemic area in eastern Indonesia, ELISA was used to measure antibody responses against merozoite proteins- MSP2, EBA-175, and PfRh2a, and to determine the association between antibody levels, malaria infection, age, and season of transmission. The results showed that antibody levels to the 3 recombinant antigens were significantly higher during the wet season, indicating increased transmission. IgG levels to MSP-2, and EBA-175 increased with age, but these remained constant for PfRh2a. The median antibody levels to these 3 antigens were higher in the falciparum-infected samples compared to the non-infected, but not statistically significant (Sennang et al., 2014).

In Haiti, where endemic-epidemic malaria occurs, serum samples were screened for antibodies against AMA-1 and MSP-1_{19} using ELISA. The results showed a seroprevalence of 23.6% to MSP-1_{19}, 21.1% to AMA-1, increased seroprevalence with age, seroconversion rate of 1.8% to MSP-1_{19}, and 1.6% to AMA-1. These findings may indicate a relatively constant state of low malaria transmission occurring, even in the absence of sustained malaria control efforts- the state of stable low transmission in this country (von Fricken et al., 2014).

In areas where elimination interventions have been implemented, serology, in combination with molecular parasitological measures, can be used to confirm areas where the infection is no longer present (Cook et al., 2012). A Malaria Indicator Survey (MIS) included a national serological survey in the Republic of Djibouti to define the extent of transmission and explore the possibilities of malaria elimination. Using ELISA, antibodies to *P. falciparum* MSP-1_{19} and AMA-1 were determined for seroprevalence and SCR. It was found out that *P. falciparum* parasite prevalence was 0.5% and *P. falciparum* seroprevalence to MSP-1_{19} and/or AMA-1 was 9.9%. Sero-prevalence significantly increased with age and the countrywide *P. falciparum* SCR was estimated to be approximately 13 new exposures per 1000 persons per year. Among the 517 individuals who were positive for *P. falciparum* antibodies, 46 were children (6.9%) under the age of five years suggesting recent malaria exposure (Noor et al., 2011).

Sri Lanka has been certified malaria-free in 2016 and a study was conducted to assess the utility of ELISA to identify changes in MTI retrospectively, in two previously endemic areas where malaria transmission is now considered virtually zero, as well as to ascertain evidence for cessation of transmission. ELISA using Pf and Pv MSP-1_{19} and AMA-1 was done to determine the seroprevalence, seroconversion, and sero-reversion rates among study subjects in a rural and an urban district. The results showed a higher proportion of the study subjects being seropositive for AMA-1 antigen in both malaria species than to MSP1−19 antigen, indicating higher immunogenicity of AMA-1. Sero-prevalence in younger age groups i.e., <10 years was significantly low for all antibodies indicating low transmission. The results suggested a low single force of transmission for falciparum malaria in the urban district but alluded to a decline of vivax malaria starting 30 years ago. It also suggested a possibility of a sub-microscopic level transmission of *P. vivax* that may have remained until recently. A drastic decline in local transmission in the rural district was evident, beginning approximately 10–15 years ago for both species, and the current rates of seroconversion being extremely low or zero, confirming the absence of active transmission. (Dewasurendra et al., 2017).

CONCLUSION AND FUTURE PERSPECTIVES

According to the Global technical strategy for malaria 2016–2030 (GTS), malaria surveillance is now a core intervention in all malaria-endemic countries and in those that have eliminated malaria but remain susceptible to its re-establishment (WHO, 2018). Microscopy of blood smears and RDTs are still the methods for malaria parasite confirmation in active and passive case detection, although their sensitivity decreases as malaria infection decrease, as in a near elimination scenario. On the other hand, sero-surveillance assays have been proven to provide key epidemiologic data for assessment of the malaria situation, the effectiveness of interventions, and monitoring the malaria program. If point-of-care (POC) testing for antimalarial antibodies can be done, sero-surveillance costs are reduced, testing is simplified, no cold chain is required for collected samples, the time to get results is reduced, and immediate feedback to individuals and communities about ongoing exposure is done, malaria surveillance can be truly made useful and sustainable (Elliott et al., 2014).

With POC malaria antibody assays accessible to malaria programs, collecting national serologic data (e.g., as part of a malaria indicator survey) in areas still in the control phase, can improve targeting of control interventions to broad areas with the highest exposure risk. In areas of lower transmission, focal surveys using these assays can allow for interventions targeted to smaller-scale hotspots at the level of villages or groups of households. In settings nearing malaria elimination, these assays can identify individuals infected in the recent past, allowing identification of geographic or demographic risk factors and ultimately, certifying that elimination has occurred. Finally, repeated evaluation of recent exposure over time can be used to assess the impact of control interventions in reducing exposure and detecting reintroduction of malaria transmission after local elimination (Helb et al., 2015).

Considering the occurrence of multiple malaria parasites and individual variations in antibody responses to multiple antigens, the search for novel and key biomarkers continues. Recently, several multiplex assays that were testing for different serological markers in the same blood sample, were developed based on the Luminex technology (Kerkhof et al., 2015; Arnold et al., 2014; Ondigo et al., 2012; Ambrosino et al., 2010; (Fernandez-Becerra et al., 2010 and microarray (Baum et al., 2015). Those biomarkers that would be identified and proven to be reliable, are then needed to be translated to high-throughput or POC assays for routine sero-surveillance (Yao et al., 2016; Rocha et al., 2017).

REFERENCES

Ambrosino, E., Dumoulin, C., Orlandi-Pradines, E., Remoue, F., Toure-Baldé, A., Tall, A., Sarr, J. B., Poinsignon, A., Sokhna, C., Puget, K., Trape, J. F., Pascual, A., Druilhe, P., Fusai, T., and Rogier, C. (2010). A multiplex assay for the simultaneous detection of antibodies against 15 *Plasmodium falciparum* and *Anopheles gambiae* saliva antigens. *Malaria Journal*, 9: 317. https://doi.org/10.1186/1475-2875-9-317.

Arnold, B. F., Priest, J. W., Hamlin, K. L., Moss, D. M., Colford, J. M., Jr, and Lammie, P. J. (2014). Serological measures of malaria transmission in Haiti: comparison of longitudinal and cross-sectional methods. *PloS One*, 9(4): e93684. https://doi.org/10.1371/journal.pone.0093684.

Baum, E., Sattabongkot, J., Sirichaisinthop, J., Kiattibutr, K., Davies, D. H., Jain, A., Lo, E., Lee, M. C., Randall, A. Z., Molina, D. M., Liang, X., Cui, L., Felgner, P. L., and Yan, G. (2015). Submicroscopic and asymptomatic *Plasmodium falciparum* and *Plasmodium vivax* infections are common in western Thailand - molecular and serological evidence. *Malaria Journal*, 14: 95. https://doi.org/10.1186/s12936-015-0611-9.

Bousema, T., Stevenson, J., Baidjoe, A., Stresman, G., Griffin, J. T., Kleinschmidt, I., Remarque, E. J., Vulule, J., Bayoh, N., Laserson, K., Desai, M., Sauerwein, R., Drakeley, C., and Cox, J. (2013). The impact of hotspot-targeted interventions on malaria transmission: study protocol for a cluster-randomized controlled trial. *Trials*, 14: 36. https://doi.org/10.1186/1745-6215-14-36.

Bretscher, M. T., Supargiyono, S., Wijayanti, M. A., Nugraheni, D., Widyastuti, A. N., Lobo, N. F., Hawley, W. A., Cook, J., and Drakeley, C. J. (2013). Measurement of *Plasmodium falciparum* transmission intensity using serological cohort data from Indonesian schoolchildren. *Malaria Journal*, 12: 21. https://doi.org/10.1186/1475-2875-12-21.

Cook, J., Reid, H., Iavro, J., Kuwahata, M., Taleo, G., Clements, A., McCarthy, J., Vallely, A., and Drakeley, C. (2010). Using serological measures to monitor changes in malaria transmission in Vanuatu. *Malaria Journal*, 9: 169. https://doi.org/10.1186/1475-2875-9-169.

Cook, J., Speybroeck, N., Sochanta, T., Somony, H., Sokny, M., Claes, F., Lemmens, K., Theisen, M., Soares, I. S., D'Alessandro, U., Coosemans, M., and Erhart, A. (2012). Sero-epidemiological evaluation of changes in *Plasmodium falciparum* and *Plasmodium vivax* transmission patterns over the rainy season in Cambodia. *Malaria Journal*, 11: 86. https://doi.org/10.1186/1475-2875-11-86.

Corran, P., Coleman, P., Riley, E., and Drakeley, C. (2007). Serology: a robust indicator of malaria transmission intensity?. *Trends in Parasitology*, 23(12): 575–582. https://doi.org/10.1016/j.pt.2007.08.023.

Cutts, J. C., Powell, R., Agius, P. A., Beeson, J. G., Simpson, J. A., and Fowkes, F. J. (2014). Immunological markers of *Plasmodium vivax* exposure and immunity: a systematic review and meta-analysis. *BMC Medicine*, 12: 150. https://doi.org/10.1186/s12916-014-0150-1.

Dewasurendra, R. L., Dias, J. N., Sepulveda, N., Gunawardena, G. S., Chandrasekharan, N., Drakeley, C., and Karunaweera, N. D. (2017). Effectiveness of a serological tool to predict malaria transmission intensity in an elimination setting. *BMC Infectious Diseases*, 17(1): 49. https://doi.org/10.1186/s12879-016-2164-0.

Elliott, S. R., Fowkes, F. J., Richards, J. S., Reiling, L., Drew, D. R., and Beeson, J. G. (2014). Research priorities for the development and implementation of serological tools for malaria surveillance. *F1000 Prime Reports*, 6: 100. https://doi.org/10.12703/P6-100.

Fernandez-Becerra, C., Sanz, S., Brucet, M., Stanisic, D. I., Alves, F. P., Camargo, E. P., Alonso, P. L., Mueller, I., and del Portillo, H. A. (2010). Naturally-acquired humoral immune responses against the N- and C-termini of the *Plasmodium vivax* MSP1 protein in endemic regions of Brazil and Papua New Guinea using a multiplex assay. *Malaria Journal*, 9: 29. https://doi.org/10.1186/1475-2875-9-29.

Fowkes, F. J., Richards, J. S., Simpson, J. A., and Beeson, J. G. (2010). The relationship between anti-merozoite antibodies and incidence of *Plasmodium falciparum* malaria: A systematic review and meta-analysis. *PLoS Medicine*, 7(1): e1000218. https://doi.org/10.1371/journal.pmed.1000218.

Helb, D. A., Tetteh, K. K., Felgner, P. L., Skinner, J., Hubbard, A., Arinaitwe, E., Mayanja-Kizza, H., Ssewanyana, I., Kamya, M. R., Beeson, J. G., Tappero, J., Smith, D. L., Crompton, P. D., Rosenthal, P. J., Dorsey, G., Drakeley, C. J., and Greenhouse, B. (2015). Novel serologic biomarkers provide accurate estimates of recent *Plasmodium falciparum* exposure for individuals and communities. *Proceedings of the National Academy of Sciences of the United States of America*, 112(32): E4438–E4447. https://doi.org/10.1073/pnas.1501705112.

Herman, L. S., Fornace, K., Phelan, J., Grigg, M. J., Anstey, N. M., William, T., Moon, R. W., Blackman, M. J., Drakeley, C. J., and Tetteh, K. (2018). Identification and validation of a novel panel of *Plasmodium knowlesi* biomarkers of serological exposure. *PLoS Neglected Tropical Diseases*, 12(6): e0006457. https://doi.org/10.1371/journal.pntd.0006457.

Kerkhof, K., Canier, L., Kim, S. Heng, S., Sochantha, T., Sovannaroth, S., Vigan-Womas, I., Coosemans, M., Sluydts, V., Ménard, D., and Durnez, L. (2015). Implementation and application of a multiplex assay to detect malaria-specific antibodies: a promising tool for assessing malaria transmission in Southeast Asian pre-elimination areas. *Malaria Journal*, 14: 338. https://doi.org/10.1186/s12936-015-0868-z.

Kim, J., Jang, J. W., Kim, J. Y., Oh, D. J., and Lim, C. S. (2016). Combined use of malaria antigen and antibody enzyme-linked immunosorbent assay for blood screening of *Plasmodium vivax* in the Republic of Korea. Medical principles and practice: *International*

Journal of the Kuwait University, Health Science Centre, 25(3): 212–218. https://doi.org/10.1159/000444144.

Müller-Sienerth, N., Shilts, J., Kadir, K. A., Yman, V., Homann, M. V., Asghar, M., Ngasala, B., Singh, B., Färnert, A., and Wright, G. J. (2020). A panel of recombinant proteins from human-infective *Plasmodium* species for serological surveillance. *Malaria Journal*, 19(1), 31. https://doi.org/10.1186/s12936-020-3111-5.

Noor, A. M., Mohamed, M. B., Mugyenyi, C. K., Osman, M. A., Guessod, H. H., Kabaria, C. W., Ahmed, I. A., Nyonda, M., Cook, J., Drakeley, C. J., Mackinnon, M. J., and Snow, R. W. (2011). Establishing the extent of malaria transmission and challenges facing pre-elimination in the Republic of Djibouti. *BMC Infectious Diseases*, 11: 121. https://doi.org/10.1186/1471-2334-11-121.

Ondigo, B. N., Hodges, J. S., Ireland, K. F., Magak, N. G., Lanar, D. E., Dutta, S., Narum, D. L., Park, G. S., Ofulla, A. V., and John, C. C. (2014). Estimation of recent and long-term malaria transmission in a population by antibody testing to multiple *Plasmodium falciparum* antigens. *The Journal of Infectious Diseases*, 210(7): 1123–1132. https://doi.org/10.1093/infdis/jiu225.

Ondigo, B. N., Park, G. S., Gose, S. O., Ho, B. M., Ochola, L. A., Ayodo, G. O., Ofulla, A. V., and John, C. C. (2012). Standardization and validation of a cytometric bead assay to assess antibodies to multiple *Plasmodium falciparum* recombinant antigens. *Malaria Journal*, 11: 427. https://doi.org/10.1186/1475-2875-11-427.

Rocha, M. V., Françoso, K. S., Lima, L. C., Camargo, T. M., Machado, R., Costa, F., Rénia, L., Nosten, F., Russell, B., Rodrigues, M. M., and Soares, I. S. (2017). Generation, characterization and immunogenicity of a novel chimeric recombinant protein based on *Plasmodium vivax* AMA-1 and MSP119. *Vaccine*, 35(18): 2463–2472. https://doi.org/10.1016/j.vaccine.2017.03.023.

Rosas-Aguirre, A., Speybroeck, N., Llanos-Cuentas, A., Rosanas-Urgell, A., Carrasco-Escobar, G., Rodriguez, H., Gamboa, D., Contreras-Mancilla, J., Alava, F., Soares, I. S., Remarque, E., D Alessandro, U., and Erhart, A. (2015). Hotspots of malaria transmission in the

Peruvian Amazon: rapid assessment through a parasitological and serological survey. *PloS One*, 10(9): e0137458. https://doi.org/10.1371/journal.pone.0137458.

Sennang, N., Rogerson, S., Wahyuni, S., Yusuf, I., and Syafruddin, D. (2014). Antibody response against three *Plasmodium falciparum* merozoite antigens in Mamuju District, West Sulawesi Province, Indonesia. *Malaria Journal*, 13: 381. https://doi.org/10.1186/1475-2875-13-381.

Stewart, L., Gosling, R., Griffin, J., Gesase, S., Campo, J., Hashim, R., Masika, P., Mosha, J., Bousema, T., Shekalaghe, S., Cook, J., Corran, P., Ghani, A., Riley, E. M., and Drakeley, C. (2009). Rapid assessment of malaria transmission using age-specific sero-conversion rates. *PloS One*, 4(6): e6083. https://doi.org/10.1371/journal.pone.0006083.

Supargiyono, S., Bretscher, M. T., Wijayanti, M. A., Sutanto, I., Nugraheni, D., Rozqie, R., Kosasih, A. A., Sulistyawati, S., Hawley, W. A., Lobo, N. F., Cook, J., and Drakeley, C. J. (2013). Seasonal changes in the antibody responses against *Plasmodium falciparum* merozoite surface antigens in areas of differing malaria endemicity in Indonesia. *Malaria Journal*, 12: 444. https://doi.org/10.1186/1475-2875-12-444.

Tanner, M., Greenwood, B., Whitty, C. J., Ansah, E. K., Price, R. N., Dondorp, A. M., von Seidlein, L., Baird, J. K., Beeson, J. G., Fowkes, F. J., Hemingway, J., Marsh, K., and Osier, F. (2015). Malaria eradication and elimination: views on how to translate a vision into reality. *BMC Medicine*, 13: 167. https://doi.org/10.1186/s12916-015-0384-6.

van den Hoogen, L. L., Bareng, P., Alves, J., Reyes, R., Macalinao, M., Rodrigues, J. M., Fernandes, J. M., Goméz, L. F., Hall, T., Singh, S. K., Fornace, K., Luchavez, J., Kitchen, A., Chiodini, P., Espino, F., Tetteh, K., Stresman, G., Sepúlveda, N., and Drakeley, C. (2020). Comparison of commercial ELISA kits to confirm the absence of transmission in malaria elimination settings. *Frontiers in Public Health*, 8: 480. https://doi.org/10.3389/fpubh.2020.00480.

von Fricken, M. E., Weppelmann, T. A., Lam, B., Eaton, W. T., Schick, L., Masse, R., Beau De Rochars, M. V., Existe, A., Larkin, J., 3rd, and Okech, B. A. (2014). Age-specific malaria seroprevalence rates: a cross-sectional analysis of malaria transmission in the Ouest and Sud-Est departments of Haiti. *Malaria Journal*, 13: 361. https://doi.org/10.1186/1475-2875-13-361.

Yao, M. X., Sun, X. D., Gao, Y. H., Cheng, Z. B., Deng, W. W., Zhang, J. J., & Wang, H. (2016). Multi-epitope chimeric antigen used as a serological marker to estimate *Plasmodium falciparum* transmission intensity in the border area of China-Myanmar. *Infectious Diseases of Poverty*, 5(1): 98. https://doi.org/10.1186/s40249-016-0194-x.

In: ELISA
Editor: Jose Ma. M. Angeles
ISBN: 978-1-68507-500-2

Chapter 5

Applications of Enzyme-Linked Immunosorbent Assays in Functional Food Development and Regulation

Deborah S. Nicdao*[1]*, Princess Alyssa S. Abid*[1,3]*,
Mia Beatriz C. Amoranto*[1]*, Joseph Q. Bechayda*[1]*,
Sharah Kristine G. Bonguit*[1,3]*,
***Calline Danica D. Gomez*[1,3]**
***and Marilen Parungao Balolong*[1,2,*]**

[1]Applied Microbiology for Health and Environment Research Group (AMHERG), University of the Philippines Manila, Manila, Philippines
[2]Department of Biology, College of Arts and Sciences, University of the Philippines Manila, Manila, Philippines
[3]Department of Biochemistry and Molecular Biology, College of Medicine, University of the Philippines Manila, Manila, Philippines

[*] Corresponding Author's E-mail: mpbalolong.up.edu.ph.

ABSTRACT

Foods are widely known to provide nutrition necessary for life and growth. They are classified according to origin, nutrients, function, and nutritive value. Recently, functional foods are gaining attention for their positive effects on human health and well-being. They are defined as ordinary foods with added components or ingredients to confer a specific health benefit other than their own nutritional value and are either industrially processed or natural foods. Examples of these additives are prebiotics, probiotics, and postbiotics, all of which contribute to the enhanced health benefits of functional foods. Because of their potential to improve immune function, functional foods provide practical and realistic adjunct therapy options for patients with metabolic-related diseases, neurocognitive disorders, cancer, and human immunodeficiency virus (HIV).

Identifying more effective interventions that may delay, if not halt, the progression of diseases requires understanding the important biological and pathogenic processes involved. Since it is quite challenging to study all these processes, focusing on specific factors and biological variables through a more targeted approach is ideal. Biological markers or biomarkers provide significant insights into the prognosis of a specific ailment as they are measurable cellular, biochemical, or molecular alterations that act as indicators of normal biological processes, pathogenic processes, and pharmacological responses to various therapeutic interventions. Hence, they play an important role in the screening, early diagnosis, and monitoring of diseases. The growing use of biomarkers in research has contributed greatly to refining the current knowledge in various medical conditions including cancer, HIV, metabolic-related complications, and neurocognitive disorders.

Since the development of functional food involves studying molecular interactions between enzymes and their inhibitors/enhancers, as well as detecting the presence/absence of biomarkers, it requires a method that could accurately detect and measure, qualitatively and/or quantitatively, the aforementioned substances even at low concentrations and in complex solutions (e.g., plasma). A prime example of which is the enzyme-linked immunosorbent assay (ELISA).

At present, the Philippines has a wide range of fermented foods that are being explored and developed to become functional foods. Traditional fermented foods are also a rich source of lactic acid bacteria (LAB), and some studies have shown the potential roles of local LAB strains in human health. Unfortunately, reports on the development and regulation of Philippine functional foods and their potential as adjunct therapy are still lacking. Therefore, utilizing ELISA as a method of detection of immunomodulation is a promising approach in developing local functional foods available for all and addressing several health concerns in the Philippines.

Keywords: ELISA, functional food, biomarkers

INTRODUCTION

There is a global paradox that in the next 20 years, the food supply will be insufficient for the growing human population and at the same time, cases of obesity and associated diseases will rapidly increase (London, 2010). In response to this, our notion of nutrition has totally evolved throughout the years. From consuming food to provide energy and nutrients for normal bodily functions and prevention of deficiencies, we are now shifting to designing foods that promote optimal health conditions and reduce the risk of developing chronic diseases (Clydesdale, 2004; Villaño, et al., 2016). The concept of natural or supplemented foods that improve immune functions is not new. Manuka honey, which was used by ancient Egyptians to treat wounds, is now being produced from farmed bees in Australia and New Zealand due to their potential in reducing infection and alleviating flu and sore throat symptoms. Physiologically active components in foods (e.g., phytochemicals and zoochemicals) are continuously being identified and studied to address existing and emerging diseases. Adding these components to our diet to create the so-called "functional foods" is a promising trend in food and nutritional science (German, 1999; Hasler, 2002; London, 2010).

To develop functional foods against various diseases, scientists should be fully knowledgeable of the biological markers associated with these diseases. Measuring these biomarkers after the administration of functional foods will objectively determine their effectiveness as adjuncts in treating or preventing the disease (Palou, Serra, and Pico, 2003). Several biomarkers have already been identified for colon cancer (i.e., colonic mucosal markers, fecal water markers, and immunological markers), cardiovascular diseases (CVDs), hypertension, obesity and other metabolic disorders, and diabetes (Hill and Peters, 2002; Rafter, 2002; Palou, Serra, and Pico, 2003; Koutelidakis and Dimou, 2016; Konstantinidi and Koutelidakis, 2019). Aside from clinical biomarkers, nutritional biomarkers are also assessed to monitor food intake and the metabolism of its dietary components. In a systematic review conducted by Li and her colleagues (2021), an extensive list of food intake biomarkers associated with fermented foods and drinks (e.g., coffee, wine, cocoa, beer, tea, cheese, and bread) was reported. Nutritional biomarkers are indicative of the bioavailability of certain nutrients and, thereby, can be utilized in analyzing one's nutritional status (Picó, et al., 2019). Moreover, the kinetics and dynamics of biomarkers in the human body – "generation, distribution, metabolism, and excretion" – should also be considered in determining the efficacy of functional foods and added ingredients (Verhagen, et al., 2004).

Various techniques have been employed for the identification and quantification of biomarkers reflecting the effects of functional food intake. To guarantee food authenticity and integrity, spectrometric methods, such as liquid chromatography coupled to mass spectrometry, tandem mass spectrometry, and high-resolution mass spectrometry, are performed as routine tests (Oh, et al., 2018; Campmajó, Núñez, and Núñez, 2019). High performance liquid chromatography (HPLC) and immunological methods, such as enzyme-linked immunosorbent assay (ELISA) and immunochromatographic strips, are also used for the rapid detection and screening of food allergens, toxins, pathogens, antibiotics, and other contaminants (Guo, et al., 2016; Kazemi and Feizy, 2020; Omar, Haddad, and Parisi, 2020). Specifically, this report focuses on the use of the different types of ELISA to develop functional foods against chronic diseases, such as cancer, HIV, metabolic-related complications, and neurocognitive disorders. An extensive list of reported biomarkers associated with the mentioned diseases are also described in this paper. Lastly, future directions on how ELISA can be utilized in functional food development and regulation in the Philippines are discussed.

Functional Food Development: Definitions and Applications

Definition of Functional Food

Foods are widely known to provide nutrition necessary for life and growth. They are classified according to origin, chemical properties, bodily function, and nutritive value. Recently, functional foods are gaining attention for their positive effect on human health and well-being. They are defined as ordinary foods with added components or ingredients to confer a specific health benefit other than their own nutritional value and are either naturally procured or industrially processed (Linares, et al., 2017). The idea of functional foods was first described in the ancient Vedic texts from India and Chinese traditional medicine (Henry, 2010). However, development of functional foods emerged in Japan during the early 1980s in the form of dietary fibers (Henry, 2010). Developed countries such as the United States, Canada, and several European countries later adopted the concept of functional foods as an alternative healthy diet to the community (Karelakis, et al., 2020). Nowadays, it is widely common for consumers to buy and consume functional foods because of the health benefits they confer.

Food products can be made functional by using any of the five approaches (Henry, 2010) – 1) removal of components known to cause damaging effects when consumed, 2) increasing the concentration of a naturally present component, 3) addition of components not normally present in foods, 4) replacing the components, usually macronutrients, with components known to have beneficial effects; and 5) increasing the stability or bioavailability of components known to have beneficial effects. Functional foods are divided into three general categories (Henry, 2010) – 1) conventional foods that contain naturally-occurring bioactive substances, such as fish (Sarojnalini and Hei, 2019), grains (Sidhu, et al., 2007), and legumes (Angeles, et al., 2021); 2) foods modified by enrichment with bioactive substances (Sheehan, et al., 2007; Figueroa and Genovese, 2019), and 3) synthesized food ingredients.

Prebiotics, Probiotics, and Postbiotics

Most functional foods are recognized to deliver probiotics. Probiotics are defined as "live microorganisms which when administered in adequate amounts confer a health benefit to the host" (Net, et al., 2014). Probiotic foods or supplements contain live beneficial bacteria or yeast that are "generally recognized as safe." The most commonly used probiotics in human nutrition include specific strains from the genera *Lactobacillus, Bifidobacterium, Enterococcus, Streptococcus, Pediococcus, Leuconostoc, Bacillus*, and *Saccharomyces* (Fijan, 2014). Probiotic strains are good ingredients for functional food development because of their ability to resist acid and bile concentrations of the gastrointestinal tract, adhere and colonize the intestinal epithelium, exclude pathogenic microorganisms, and produce antimicrobial peptides (Syngai, et al., 2016). Moreover, several lactic acid bacteria (LAB) considered as probiotics have immunomodulatory properties (Tsai, et al., 2012; Amoranto and Balolong, 2020). An example of a functional food that delivers probiotics is kefir, a fermented dairy product with a relatively stable and specific balance of LAB and yeast (Guzel-Seydim, et al., 2011). *Lactobacillus* spp. strains from kefir were found to exert health benefits, especially on cholesterol metabolism, diabetes, and tumor suppression (Slattery et al., 2019).

Meanwhile, prebiotics are also added to functional foods to confer health benefits to the host (Peng, et al., 2020). Prebiotics are non-digestible, short-chain carbohydrates that are selectively fermented in the gut (Al-Sheraji, et al., 2013). They should be able to resist acid and bile concentrations, be fermented by the gut bacteria, and enhance the activity of beneficial microorganisms such as probiotics (Al-Sheraji, et al., 2013). Examples of prebiotics that have been recently utilized to develop functional foods are inulin, B-glucan, xylitose, and glutamine (Peng, et al., 2020). Prebiotic-enriched functional foods were observed to prevent the onset of diet-related diseases (Florowska, et al., 2016), reduce cancer risk (Samanta, 2020), and alleviate chronic infections (Younis, et al., 2015). Interestingly, recent studies have shown that the synergy between prebiotics and probiotics helps in maintaining balance in the gut microbiome.

Recently, postbiotics are gaining attention as an ingredient for functional foods. Postbiotics are by-products and soluble factors secreted by live bacteria or released after cell lysis (Aguilar-Toalá, et al., 2018). There are no reported risks of bacterial translocation and transfer of antibiotic resistance genes when utilizing postbiotics as components of functional foods (Nataraj, et al., 2020; Rad, et al., 2020). Moreover, postbiotics have favorable absorption, metabolism, and excretion abilities, which could elicit host responses (Aguilar-Toalá, et al., 2018).

Promising Developments for Health and Wellness: Cancer, HIV, Metabolic-Related Complications, Neurocognitive Disorders

Functional foods are being perceived to contribute to a healthy and balanced diet, and consumers believe that they reduce the risks of health problems (Karelakis, et al., 2020). Due to the increase in popularity of functional foods, promising food products for health and wellness are being developed. Several studies have shown the effects of functional foods in health problems, such as cancer, HIV, metabolic-related complications, and neurocognitive disorders. According to WHO, cancer is the second leading cause of death globally, with approximately 10 million deaths in 2020 (World Health Organization, 2021). Because of the threat caused by cancers, several research studies are focused on cancer treatment. Current treatments for cancer depend on the type of cancer, stage at diagnosis, and the person's overall health. As of writing, there are very few studies on the effects of functional food on cancer-related cases. In mouse models, lycopene has shown anti-carcinogenic roles in mammary gland, liver, skin, and lungs, and it also prevented the development of aberrant crypt foci in rat colon (Nishino, et al., 2019). It has also been demonstrated that people who consume diets high in omega-3 fatty acids may experience a lower prevalence of some types of cancer (Takezaki, et al., 2003). Phenolic flavonoids, such as green tea polyphenols and epigallocatechin-3-gallate, also act as anticancer agents by activating the transcription system (Xu, et al., 2005).

Metabolic syndrome (MetS) is also a growing health concern worldwide. It is characterized by several metabolic abnormalities, such as hypertension, insulin resistance, obesity, dyslipidemia, increased oxidative stress, and increased pro-inflammatory state (Eckel, et al., 2005). Current treatment of MetS aims to reduce the risk of developing further complications, and treatment includes lifestyle change, exercise, and proper diet together with medications to reduce blood pressure, cholesterol, and blood sugar. Interestingly, the administration of functional foods has also been proven effective in the management and improvement of these chronic diseases. Several functional foods have been reported to reduce cholesterol absorption (Biörklund, et al., 2005; Chen, et al., 2008) and insulin sensitivity (Sacks, et al., 2001; Mirmiran, 2014), decrease inflammation (Luvián-Morales, et al., 2021), and induce weight loss (van der Made, et al., 2012).

Neurocognitive diseases are characterized by impaired mental functions, memory loss, confusion, anxiety, changes in behavior, and having trouble in performing daily tasks (Legg, 2018). Neurocognitive diseases are more prevalent in older individuals as aging is considered one of its greatest risk factors. Current treatment includes antibiotics, occupational and physical therapy, pain medications, and surgery (Legg, 2018). However, some neurocognitive disorders, such as dementia and Alzheimer's disease, have no cure, compared to a concussion or an infection that are temporary and curable. Functional food aims to supplement diet, as the diet affects adult neurogenesis. Studies on dietary factors affecting the brain found compounds that can improve cognition because of their antioxidant and anti-inflammatory properties (Poulose, et al., 2017). Vitamins B_9 and E, omega-3 polyunsaturated fatty acids, and non-nutrient phytochemicals also improved adult neurogenesis (Poulose, et al., 2017). Commonly found in fatty fish and walnuts, omega-3 fatty acids are necessary in the structure and function of the brain. However, omega-3 supplementation only shows effect on cognition in certain compromised groups and not in the healthy adult population (Poulose, et al., 2017). Polyphenol-rich foods like blueberries have shown to increase neurogenesis (Casadesus, et al., 2004). Interestingly, several functional foods are reported to be promising therapies against Alzheimer's Disease (Atlante, et al., 2020). These studies show that functional foods are promising practical food interventions for patients with neurocognitive disorders, as food insecurity may accelerate cognitive decline (Na, et al., 2020).

The human immunodeficiency virus (HIV) is one of the most studied immune diseases for probiotic intervention. In 2019, the number of people living with HIV reached 38 million with 1.7 million recorded new infections (World Health Organization, 2020). People living with HIV have weakened immune systems which makes them more susceptible to illness and complications, such as diarrhea and dyslipidemia. Currently, there is no effective cure for HIV, but taking proper medications can control the viral load. Antiretroviral therapy (ART) is a combination of HIV medicine that can reduce viral load to an undetectable level and reduce the risk of HIV transmission. However, ART may be limited in other countries, so alternative therapies should be developed. A novel hypothesis suggests that probiotics protect the gastrointestinal tract, reducing the incidence of diarrhea and postponing HIV progression to acquired immunodeficiency syndrome (AIDS). Yogurt fortified with *Moringa oleifera* leaves and probiotic *L. rhamnosus* GR-1 was developed to provide a new functional food for populations with high HIV infection rates (Van Tienen, et al., 2011). LAB supplementation in ART-stable HIV patients showed a reduction in coagulation and inflammation markers and improvement in the gut microbiota composition after eight weeks of probiotic intervention (Stiksrud, et al., 2015). Meanwhile, LAB supplementation in patients with HIV not receiving ART revealed that consumption of probiotics twice daily for six months significantly prevented $CD4^+$ count reduction (Emadi-Koochack, et. al. 2019). These studies show that functional foods are promising therapies for people living with HIV and encourage the development of more functional foods targeting HIV.

Because of the potential of functional foods in improving health and wellness in several diseases, they now provide a practical and realistic adjunct therapy option for patients with metabolic-related diseases, neurocognitive disorders, cancer, and HIV. There is a need for continuous development of functional foods with novel health benefits, and elucidating their mechanisms of action is also of importance.

Biomarkers and Biological Disease Models for Health

Biomarkers for Cancer

Cancer is one of the leading causes of death worldwide with 19 million new cases and 9.95 million deaths in 2020. WHO predicts that this number will balloon to 30.2 million new cases and 16.3 million deaths by year 2040. Cancer is defined as a complex and heterogeneous group of diseases characterized by an uncontrollable rate of cell growth and proliferation. These cells may spread to other parts of the body forming visible growth called tumors or lumps of tissues (National Cancer Institute, 2021). Tumors can be malignant or benign. Malignant or cancerous tumors invade other tissues and can travel distant places to form new tumors (a process called metastasis). Benign tumors, on the other hand, do not spread into other tissues, and they usually do not grow back when removed, whereas malignant tumors sometimes do. However, large benign tumors may still be life-threatening, such as when they are formed in the brain. There are more than 100 different types of cancer on which they are usually named after the organs or tissues where the tumors form. However, there are also other types of cancer, such as leukemia, cancer of the blood, which do not present tumors (National Cancer Institute, 2021).

A biological marker or biomarker is any measurable substance, structure, or product from a physiological process that influence or predict the incidence or outcome of the disease (Mayeux, 2004). These are usually found in the blood or urine for easier and a less invasive method of collection of samples. In cancer, these can be biomolecules (protein, peptide, DNA, or RNA) that are products of tumor cells or of the surrounding tissues as a response to the tumor. The use of cancer biomarkers (CBs) aims to guide physicians in making assessments in different clinical settings based on the state of the patient. CBs can help in screening, estimating the risk of disease, distinguishing benign from malignant, or a different type of malignancy, determining prognosis, monitoring the progression of the disease, and predicting recurrence or one's potential response to therapy (Henry and Hayes, 2012; Goossens, et al., 2015; Kamel and Al-Amodi, 2016).

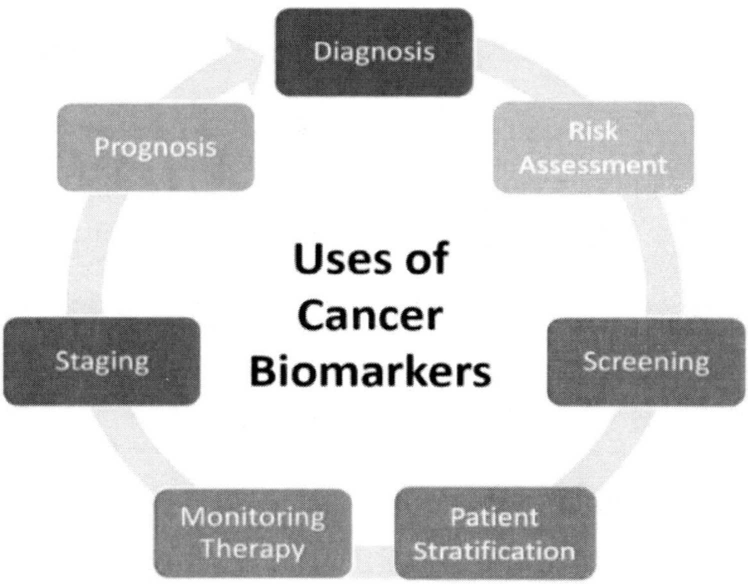

Figure 1. Clinical utility and uses of cancer biomarkers.
(Modified figure of Kamel and Al-Amodi, 2016).

Table 1. List of some FDA-approved cancer biomarkers currently used in clinical practice

Organ Specificity/ Cancer Type	Biomarker	Basis for considering as a biomarker	Application/ Uses	References
Prostate/ BPH	Prostate-specific antigen (PSA)	A protein produced by both cancerous and noncancerous tissue in the prostate	Screening, diagnosis, and monitoring therapy	Bjartell, 2013; Vickers and Brewster, 2012
Prostate	Prostate cancer antigen 3 (PCA3)	A prostate specific, nonprotein coding RNA that is significantly overexpressed in prostate cancer	Screening and prognosis	van Gils, et al., 2007
Ovarian	BRCA1 and 2 gene mutations	Breast cancer genes that produce proteins to repair damaged DNA. Mutations cause BRCA1 or BRCA1 proteins to grow out of control and may lead to cancer	Screening	Easton, et al., 1995; Hall, et al., 1990
	Cancer antigen 125 (CA125)	A protein produced by a variety of cells, particularly ovarian cancer cells.	Prognosis It is used for patients that are high-risk for ovarian cancer or when monitoring treatment or recurrence.	van Nagell, et al., 2000
Colorectal	Multitarget stool DNA test (aberrantly methylated BMP3 and NDRG4 promoter, mutant KRAS, and β-actin)	Causative mutations	Screening	Imperiale, et al., 2014

Organ Specificity/ Cancer Type	Biomarker	Basis for considering as a biomarker	Application/ Uses	References
	KRAS	Mutations activate RAS-RAF-MEK pathway and resistance to EGFR therapy	Monitoring therapy	Van Cutsem, et al., 2009
Colorectal/ hepatic	Carcino-embryonic antigen (CEA)	CEA is a complex glycoprotein produced by 90% of colorectal cancers and contributes to the malignant characteristics of a tumor	Monitoring therapy, prognosis, detecting recurrence, screening for hepatic metastases	Koizumi et al., 1992; Matsuoka, 1990; Von Kleist, 1986
Gastro-intestinal Stromal Tumor (GIST)	KIT	Mutation leads to constitutional activation	Monitoring therapy	Dematteo et al., 2009
Breast	Cancer antigen 15.3 (CA 15-3)	CA 15-3 is slightly elevated in <50% of patients with early localized breast cancer but is elevated in about 80% of those with metastatic breast cancer	Monitoring therapy	Bast, et al., 2001; Molina, et al., 2005
	Estrogen, progesterone receptors (ER and PgR)	Suggests sensitivity to endocrine therapy	Stratification/ select patients for endocrine therapy	McGuire, et al., 1997; Coolleoni, et al., 2004; Paik, et al., 2006
	HER2	Oncogene overexpression	Monitoring trastuzumab therapy	Varadan, et al., 2016; Ross, et al., 2009; Romond, et al., 2005

Table 1. (Continued)

Organ Specificity/ Cancer Type	Biomarker	Basis for considering as a biomarker	Application/ Uses	References
Breast	Cancer antigen 27.29 (CA 27.29)	CA 27.29 is primarily used to monitor metastatic breast cancer and to detect recurrence	Monitoring therapy, recurrence	Jensen, et al., 1991
	21 gene recurrence score	A multigene assay designed to measure the expression of 21 genes – 16 cancer-related genes and 5 endogenous references in breast tumors. s	Prognosis This assay is used to predict chemotherapy benefit and guide adjuvant treatment decision in HR+/HER2-/node- breast cancer patient	Paik, et al., 2004
Testicular	Human chorionic gonadotropin-β (HCG-β)	Cancerous cells from germ cell tumors of the testis transform into syncytio-trophoblasts (a normal component of the placenta) and secrete HCG.	Diagnosis, staging, detecting recurrence, monitoring therapy	Kurtzman and Wilson, 2001
Hepato-cellular carcinoma	Alfa-fetoprotein (AFP)	AFP is a protein normally produced in the liver of a developing fetus. However, it is used as a tumor marker in adults as healthy adults should have very low levels of AFP.	Diagnosis, detecting recurrence, monitoring therapy	Chen, et al., 2010; Johnson and Williams, 1980; Minami, et al., 2015

Organ Specificity/ Cancer Type	**Biomarker**	**Basis for considering as a biomarker**	**Application/ Uses**	**References**
Medullary carcinoma of thyroid	Calcitonin	It is one of the hormones produced by the thyroid, but its level is increased in patients with medullary thyroid cancer.	Diagnosis and monitoring therapy	Bae, et al., 2015; van Veelen, et al., 2009
Thyroid	Thyroglobulin (TG)	TG is a glycoprotein produced by the thyroid follicular cells. Increased level of TG during postoperative period may be due to residual thyroid tissue.	Monitoring, detecting recurrence after surgery	Mazzaferri, et al., 2003
Pancreatic	Carbohydrate antigen (CA) 19-9	CA 19-9 is a type of antigen released by pancreatic cancer cells	Monitoring therapy	Ludwig and Weinstein, 2005
Bladder	Nuclear matrix protein (NMP-22)	A specific nuclear matrix protein able to detect a bladder tumor via voided urine	Screening, monitoring and prognosis	Kibar, et al., 2006
Hemato-logical, chronic myeloid leukemia	BCR-ABL	Balanced t(9,22) leading to the formation of a constitutively active tyrosine kinase	Monitoring therapy	Kantajiran, et al., 2002
Lung, NSCLC Lung, NSCLC	EGFR (HER1)	Mutations in tyrosine kinase domain	Monitoring therapy	Pao and Girard, 2011
	ALK	Inversion in chromosome 2 leads to EML4-ALK fusion oncogene	Monitoring therapy	Shaw, et al., 2013

Table 1. (Continued)

Organ Specificity/ Cancer Type	Biomarker	Basis for considering as a biomarker	Application/ Uses	References
Skin, Melanoma	BRAF V600	80-90% V600E mutation, a downstream mediator of RAS, leads to downstream activation of MEK and ERK	Monitoring therapy	Chapman, et al., 2011

CBs are said to be classified into different categories based on their usage. To cite a few examples, predictive biomarkers indicate how a patient will respond to specific therapeutic interventions, such as positivity/activation of *HER2* that shows that the patient is responding to trastuzumab treatment in breast cancer (Romond, et al., 2005; Ross, et al., 2009; Varadan, et al., 2016). Another example is *KRAS*-activating mutations that indicate resistance to EGFR inhibitors (i.e., cetuximab) in the treatment of colorectal cancer (Van Cutsem, et al., 2009). Another category of CBs is the prognostic biomarkers that aim to inform physicians of possible cancer recurrence or disease progression in the future; prognostic biomarkers are not necessarily linked to specific treatments. An example of this is the 21-gene recurrence score, which is used to check for breast cancer recurrence and for the overall survival in tamoxifen-treated breast cancer (Paik, et al., 2004). Another class of CBs is the diagnostic biomarkers that aims to identify or screen for a specific type of cancer. An example of this is testing for stool cancer DNA, which is implemented in colorectal cancer surveillance (Imperiale, et al., 2014).

Table 1 lists some of the different CBs currently approved by the Food and Drug Administration (FDA) that are used in the diagnosis and prognosis of the different types of cancer. However, it should be noted that there is no perfect or ideal biomarker, and the use of these biomolecules should be combined with other tests when assessing the stage of the disease.

Biomarkers for HIV

With over 37.6 million people affected globally by HIV in 2020, HIV infection still remains a global health concern. Within the same year, 1.5 million people have also been newly infected with the virus (UNAIDS, 2020). Briefly, the HIV life cycle begins with the attachment of the HIV particle to the host cell membrane where interactions in the membrane components induce fusion between the cell and the viral membrane. Consequently, the contents of the virion are transferred to the cell cytoplasm, wherein viral DNA can access and enter the nucleus, become integrated into the host chromosome, and trigger the release of HIV particles in the body (Maeda, et al., 2019). This makes HIV infection dangerous to acquire as it rapidly disseminates in the body and at the same time primarily targets the immune system, thereby inhibiting the body from controlling the viral progression.

Fortunately, the development of antiretroviral drugs that specifically controls HIV progression has substantially decreased the burden of AIDS-related deaths over time by suppressing viral replication and reducing plasma viral load (Trickey, et al., 2017). Combined ART (cART) or the combination of these antiretroviral drugs has been implemented as one of the common methods used to control HIV progression in affected patients. The use of cART as a therapeutic strategy has evidently contributed to the reconstitution of the immune system of people living with HIV. However, the long-term dependence of HIV patients to this therapy have also initiated adverse effects (Maeda, et al., 2019). Hence, the search for reliable biomarkers that could help predict disease progression in acute HIV is also relevant in managing the infection.

Considering that HIV disease progression varies among individuals, it is still a current challenge to identify dependable biomarkers for the disease (Table 2). However, $CD4^+$ T-cell counts have shown a strong correlation with HIV progression in numerous studies. This is also because one of the characteristic features of acute HIV infection is the rapid loss of $CD4^+$ T-cells (Maartens, et al., 2014; Rutstein, et al., 2017).

Because HIV infection mainly targets the immune function of the affected individual, it is only expected that it targets cells that are crucial in regulating the patient's immune response to different pathogens. $CD4^+$ T-cells participate in this immune response through a variety of mechanisms, including activation of cells of the immune system, secretion of cytokines, and suppression of immune reaction (Luckheeram, et al., 2012). Therefore, it has been considered as one of the reliable biomarkers for acute HIV infection along with plasma viral load levels. These two markers have also been used in clinical settings to help monitor HIV progression (Turk, et al., 2017).

Table 2. Common biomarkers associated with HIV infection

Commonly associated biomarkers	Basis for considering as a biomarker	References
$CD4^+$ T-cells	Characteristic feature of HIV infection is the depletion of $CD4^+$-T cells	Maartens, et al., 2014; Rutstein, et al., 2017
Tumor necrosis factor alpha (TNF-α)	Strongly correlated with HIV viral load set point	Vaidya, et al., 2014
	An immunoregulatory cytokine expressed when HIV-1 gp120 triggers essential signaling pathways (calcium signaling pathway)	Planès, et al., 2018
Interleukin-6 (IL-6)	Plays a huge role in both pro- and anti-inflammatory effects thereby making it an important cytokine in the activation and regulation of the immune response, especially in viral infections	Velazquez-Salinas, et al., 2019
	High levels observed in HIV-infected patients with low CD4-recovery	Rosado-Sanchez, et al., 2017
Interferon gamma-induced protein 10 (IP-10)	Plasma chemokine present during early or acute HIV-1 infection and manifested strongest correlation with HIV-1 viral load	Simmons, et al., 2013
	Showed the most significant increase in acute HIV infection and exhibited validation of model prediction power	Pastor, et al., 2017

Similar to other viral infections, acute HIV infection also induces the production of antiviral and proinflammatory biomarkers. This so-called "cytokine storm" was observed to be initiated within the first few weeks following detectable viremia in HIV cases. This state of immune activation occurs at a critical time in acute HIV infection, thereby making these biomarker pathways a crucial area of study (Teigler, et al., 2018). A few of the biomarkers involved in this event are briefly described and enumerated in Table 2. Promising results may have been exhibited in these studies, but the clinical implications and diagnostic impact of these biomarkers still require further investigation.

Efforts directed towards providing insight in controlling HIV infection does not only focus on the biomarkers directly associated with the disease but also to those that are associated with its numerous comorbidities. Despite the undeniable success of cART in controlling HIV viral load, the characteristic disruption of the immune system brought about by acquiring HIV also gives way for other infections or adverse conditions to persist. Similarly, it has been observed that people living with HIV also have an increased risk in developing diseases of aging and a higher tendency for developing CVDs (Vos et al., 2016; So-Armah and Freiberg, 2019). These will be further elaborated in the next section.

Biomarkers for Metabolic-Related Complications

MetS is a cluster of metabolic derangements that typically include hypertension, obesity, and insulin resistance, among many others. MetS has been long associated with the increased risk of developing CVD, specifically, its pathogenesis has been found to play a significant role in the final inflammation pathway leading to CVD (Rochlani, et al., 2017). The development of these CVDs has gained considerable attention due to it being one of the leading causes of death worldwide even at the present day. Similarly, CVDs have also been considered a metabolic comorbidity in a variety of diseases, including HIV infection. As such, a number of scientific studies focus on the investigation of biomarkers that could assist in the risk assessment and prediction of MetS.

As critical regulators of cardiovascular function, the role of microRNAs (miRNAs) as biomarkers for cardiovascular risk has been investigated in multiple studies (Table 3). Specific circulating miRNAs have been found to have a correlation with different pathological conditions due to their being regulators of post-translational gene expression. Apart from their significant role in the proper functioning of different biological processes, they are also relatively stable, making them ideal candidates for biomarkers in a variety of relevant health conditions (Roderburg and Luedde, 2014; O'Brien, et al., 2018).

Proteins such as troponin also play a significant role as biomarkers for many CVDs, commonly with myocardial infarctions. Cardiac troponin (cTn) has been regarded as the gold standard for the detection of myocyte necrosis — a pathological hallmark of acute myocardial infarction (AMI) (Wang, et al., 2019). Troponin primarily participates in both skeletal and cardiac muscle contraction. It participates in facilitating muscle contraction by acting as an "on-off switching mechanism" of the cyclic attachment and detachment of the myosin cross bridges along the actin-based thin filaments. These events lead to contraction of the muscle in the body. It is a complex of three subunits: I, C, and T that functions to inhibit actomyosin ATPase, bind Ca^{2+}, and link the complex to tropomyosin, respectively. Due to the reliability of troponin as a biological marker for cardiovascular conditions, assays specific for the protein have been developed and improved over time for the early detection of myocardial ischemia (Yang, et al., 2014; Park, et al., 2017). This significantly contributed to the development of effective and targeted therapies for the disease.

Markers of inflammation are also typically associated with health conditions relevant to cardiovascular risk. Being an inflammatory condition, CVDs also trigger the release of proinflammatory biomarkers during the onset of the disease. Similarly, an increase in the number of reactive oxygen species (ROS) in the body is also observed.

Table 3. Common biomarkers associated with metabolic diseases

Health condition	Commonly associated biomarkers	Basis for considering as a biomarker	References
Heart failure	miRNAs (HF-specific panel of circulating miRNAs in the study: miR-18a-5p, miR-26b-5p, miR-27a-3p, miR-30e-5p, miR-106a-5p, miR-199a-3p, and miR-652-3p)	Critical regulators of cardiovascular function and are relatively stable Decreasing levels of circulating miRNAs have been associated with increasing acuity of heart failure	Zhou, et al., 2018 Ovchinnikova, et al., 2016
Vascular aging (HIV-induced and antiretroviral therapy-induced)	miR-34a	Observed to significantly increase in HIV-infected and antiretroviral agents-treated vessels; a promising biomarker for organ aging specifically of the impaired functions of vascular endothelial cells of people with vascular disease	Zhan, et al., 2016
Acute myocardial infarction (AMI)	Cardiac troponins (cTn)	A fundamental unit for muscle contraction both in heart and skeletal muscles but have isoforms exhibiting great specificity and sensitivity for cardiac muscle and myocardial injury; assays specific for cTn considered the gold-standard biomarker for acute myocardial necrosis detection (hallmark of AMI)	Park, et al., 2017; Martin, et al., 2016
Cardiovascular diseases (general)	Reactive Oxygen Species (ROS)	Initiates oxidative stress and endothelial dysfunction and has been regarded a hallmark of different cardiovascular events	Vona, et al., 2019 Daiber, et al., 2019

Table 3. (Continued)

Health condition	Commonly associated biomarkers	Basis for considering as a biomarker	References
Atherothrombosis	C-reactive protein (CRP)	A downstream biomarker of elevated IL-6, IL-1, and TNF-α; inhibits nitric oxide (NO) thereby increasing the production of endothelin-1 production leading to impairment of endothelial-dependent vascular relaxation; contributes to hypertension and endothelial dysfunction	Badimon, et al., 2018
Myocardial infarction and coronary artery disease (CAD)	B-type natriuretic peptide (BNP)	Produced in the myocardium during wall stress and participates in smooth muscle relaxation and endothelial permeability; assesses cardiac strain and function	Kotecha, et al., 2019

ROS contribute greatly to the development of oxidative stress, which generally plays a huge role in cardiovascular conditions and conditions that are associated with inflammatory processes. Oxidative stress is initiated when the production of ROS and reactive nitrogen species (RNS) exceeds the capacity and expression of antioxidant systems of the affected individual (Vona, et al., 2019). This then leads to the development of a broad variety of health conditions including cardiovascular dysfunction. Likewise, C-reactive protein (CRP), a downstream biomarker of elevated inflammatory cytokines, IL-1, IL-6, and TNF-α, also participate in the pathogenesis of CVD, specifically atherosclerosis and atherothrombosis (Badimon, et al., 2018). These are only a few of the many biomarkers associated with CVD pathogenesis (Table 3). However, it is certain that the continued search for reliable biomarkers that can be clinically used at the present day all yearn for the development of more therapeutic strategies to combat CVD.

Biomarkers for Neurocognitive Disorders

The cluster of neurocognitive disorders is characterized in the Diagnostic and Statistical Manual of Mental Disorders (DSM-V) as the presence of cognitive deficits. The term 'cognitive' refers to thought and multiple related processes, whereas the term 'neurocognitive' was applied to emphasize that the disrupted levels of neural substrates lead to symptoms, and that such disruption can be reliably measured (Sachdev, et al., 2014). DSM-V also emphasized that the disorders classified under this category are those that are "acquired rather than developmental" – impaired cognition is due to a decline from a previously attained level of functioning and has not been present since birth or early life (Ganguli, et al., 2011). The manual lists six principal domains of cognitive function: complex attention, executive function, learning and memory, language, perpetual-motor function, and social cognition (Figure 2), all of which can be objectively assessed to be likely affected in patients diagnosed with a neurocognitive disorder (Sachdev, et al., 2014).

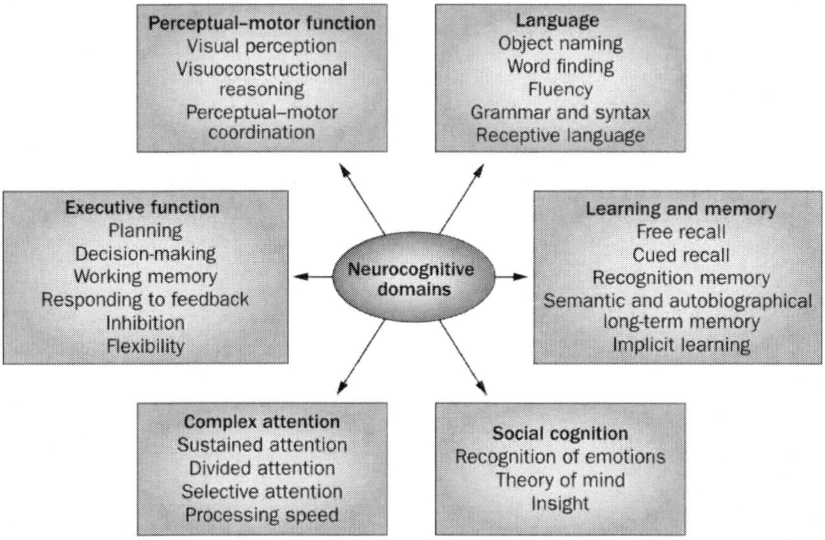

Figure 2. The neurocognitive domains. The six key domains of cognitive function defined by DSM-V (Sachdev, et al., 2014).

This cluster of disorders comprises three syndromes: delirium, major neurocognitive disorders, and the mild neurocognitive disorders, each with a range of possible etiologies. Delirium is characterized by a disturbance in attention (reduced ability to direct, sustain, and shift focus), therefore likely to have reduced orientation to their environment, and at times to oneself. This has sometimes been referred to as 'reduced level of consciousness' or 'confusional state' (Joshi, et al., 2012). It can be caused by an underlying medical condition, substance intoxication or withdrawal, exposure to toxins, or a combination of these factors (American Psychiatric Association, 2013). Major neurocognitive disorders, known previously as dementia, is defined by the DSM-V as a decline in one's mental ability, in one or more of the key cognitive domains previously highlighted, severe enough to interfere with independence and daily life. A person with mild neurocognitive disorder, on the other hand, may still be able to retain the ability to be independent. DSM-V's criteria for major and mild neurocognitive disorders are parallel with each other and only differ in the criterion of independent functioning. This key distinction between the mild and major neurocognitive disorders will rely on the reports by the individual and/or a family member, and the judgment from the clinician (Petersen, et al., 2009; Sachdev, et al., 2014).

Mild and major neurocognitive disorders are subtyped according to etiology (potential causative factors). Examples of causative disorders commonly evident in many patients with neurocognitive disorders are Parkinson's disease, Huntington disease, traumatic brain injury, HIV infection or AIDS, and stroke. Some patients manifest cognitive impairment first, and the etiology is later revealed, such as in Alzheimer's disease, cerebrovascular disease, frontotemporal lobar degeneration, and Lewy body disease. Assigning the causative factor in mild patients may be difficult and is usually left unspecified until more evidence is later presented (Ganguli, et al., 2011; Sachdev, et al., 2014). The DSM-V has a list of criteria on how to assign an etiology, and one of them is by an exclusionary criterion that the disturbance is not explained by another etiology. This includes a rigorous process of considering all possible etiologies to arrive at the most likely cause. As such, requiring additional requirements such as the use of biomarkers may be done to speed up the process and ensure a greater degree of specificity. The use of biomarkers in helping diagnose disorders associated with cognitive dysfunction has been recently recognized with some being specific enough to be added into the disease criteria. With the use of biomarkers, some may be detected early during the mild stage or pre-clinical form of the disease and may help prevent further decline (Sachdev, et al., 2014; Sonnen, et al., 2008). Some biomarkers used in neurocognitive disease research are listed in Table 4.

However, most of the biomarkers listed above are still under investigation and are yet to be validated for clinical use. Moreover, biomarkers for neurocognitive disorders are often found in cerebrospinal fluid (CSF) samples, while some biomarkers can also be derived from peripheral fluids including plasma, serum, or urine. Sample collection via the latter routes is considered more desirable and convenient but studies remain to be inconsistent and inconclusive (Sonnen, et al., 2008; McDonough and Allen, 2019). More studies regarding the blood-brain barrier leakage in these diseases must also be conducted. In addition to this, findings from one disorder usually overlap with another group of patients of another disorder as seen in the table above. Some studies also present inconsistent results depending on the age group of the same disorder. Researchers also propose that a panel of biomarkers be used instead of only using one biomarker (Sonnen, et al., 2008; Lilamand, et al., 2020; Llorens, et al., 2020). These are just some factors that make it challenging for researchers to standardize the use of biomarkers for neurocognitive disorders. Thus, the main goal in biomarker research for neurocognitive disorders should be to evaluate the usefulness of a set of standardized biomarkers, to be used together with risk factor assessment tests, neuropsychiatric tests, and structural and functional neuroimaging to help patients and physicians alike in the management of the disease.

Table 4. List of biomarkers used in research studies involving neurocognitive diseases

Disorder	Biomarker	Basis	References
Alzheimer disease (AD)	Amyloid precursor protein gene (APP) or presenilin genes (PSEN1 and PSEN2)	Causative mutations in the APP or presenilin genes	Albert, et al., 2011; Dubois, et al., 2007; Buerger, et al., 2006
	Reduced CSF levels of amyloid-β_{42} (AB42)	Amyloid deposition in the brain	
	Elevated levels of phosphorylated tau and total tau in the cerebrospinal fluid	Increased concentrations of tau species by one or two orders of magnitude compared to control level reflects NFT and neuronal death	
Frontotemporal lobar degeneration	C9ORF72 Microtubule-associated protein tau (MAPT) Granulins (GRN) TAR DNA-binding protein 43 (TDP-43) Transitional endoplasmic reticulum ATPase (VCP) Charged multivesicular body protein 2b (CHMP2B) FUS protein (FUS)	Mutations in the specified genes causing changes in the brain structure	Gorno-Tempini, et al., 2011; Rascovsky, et al., 2011
Huntington disease	Huntingtin gene (HTT)	Expansion of the HTT gene with 36 or more repeats of the CAG trinucleotide	Montine, et al., 1999
	Elevated CSF F2-isopropanes	Free radical lipid peroxidation of arachidonate due to oxidative stress	

Disorder	Biomarker	Basis	References
Lewy body disease (LBD)	Neuroimaging markers: low striatal dopamine transporter uptake on single-photon emission CT (SPECT) or PET	Low uptake means fluctuation in the attention and awareness of the patient	McKeith, et al., 2005
	Elevated CSF tau-P181	Elevated tau (microtubule-associated protein) phosphorylated at threonine 181 was able to discriminate LBD from AD	Vanderstichele, et al., 2006; Hampel, et al., 2004
Parkinson disease with dementia	Elevated CSF α-synuclein or plasma oligomers of α-synuclein	Neuronal protein that regulates synaptic vesicle trafficking and subsequent neurotransmitter release	El-Agnaf, et al., 2006; Tokuda, et al., 2006
Prion disease	Prion protein gene (PRNP)	Causative mutation	Wadsworth, et al., 2003
	Presence of 14-3-3 proteins in the CSF, increased CSF levels of tau and phosphorylated tau proteins, neurofilament light chain (NfL), and α-synuclein	Indicative of recent or ongoing rapid neuronal damage NfL is a neuronal cytoskeleton component released when there is neuronal damage	Leitão, et al., 2016; Lattanzio, et al., 2017; Bridel, et al., 2019; Llorens, et al., 2017
Vascular cognitive impairment	Increased CSF concentrations of phosphorylated tau species, combined with normal AB_{42} concentrations, increased neurofilament protein, increased CSF:serum albumin ratio increased CSF sulfatide and/or metalloproteinase concentrations	Assessment of damage from small vessel disease or of presumed white matter ischemic injury	Andreasen, et al., 2001; Sjogren, et al., 2001; Wallin and Sjogren, 2001; Fredman, et al., 1992; Adair, et al., 2004

THE ROLE OF ELISA IN FUNCTIONAL FOOD DEVELOPMENT

Definition of ELISA

Since the development of functional food involves studying molecular interactions between enzymes and their inhibitors/enhancers, as well as detecting the presence/absence of biomarkers, it requires a method that could accurately detect and measure, qualitatively and/or quantitatively, the aforementioned substances even at low concentrations and in complex solutions (e.g., plasma). Radioimmunoassay (RIA), which incorporate radioisotopes, are initially used for this purpose. However, due to safety concerns associated with radioactivity, this method has been gradually replaced by a modified immunoassay called ELISA . It is a plate-based technique that utilizes enzyme-labeled antigen/antibody to amplify antigen-antibody reactions and produce quantifiable results (Aydin, 2015; Sakamoto, et al., 2017).

Based on the immunological response of an antibody specifically binding to its antigen, ELISA can detect very small amounts of antigens, such as peptides, proteins, nucleic acids, hormones, secondary metabolites, or even antibodies in fluid samples. Analysis requires several reagents – antigen (analyte), antibodies for coating and detection, blocking agent, wash buffer, and chromogenic/fluorogenic substrate – which all depends on the type of ELISA (Engvall, 2010; Shah and Maghsoudlou, 2016; Konstantinou, 2017). In general, there are five main steps to perform ELISA, as detailed below:

1) Coating
   The antigen in the sample is immobilized on a solid phase, usually a 96-well microtiter plate made up of rigid polystyrene, polyvinyl chloride, or polypropylene (Sakamoto et al., 2017). The solid phase should strongly adsorb the antigen and prevent

other components from binding to it (Yalow and Berson, 1960; Engvall, 2010).

2) Blocking

To prevent non-specific binding of the reactants in the microtiter plates, the residual binding capacity of the wells should be eliminated using a blocking agent. This solution is a mixture of unrelated proteins and other compounds that passively adsorb to the unoccupied binding sites of the wells. Typically, bovine serum albumin (BSA) is used for this purpose, although other substances such as non-fat dry milk, whole serum, glycoprotein solutions, and non-protein blocking buffers can also be added. Another advantage of adding blocking agents is to improve sensitivity by stabilizing the coated protein, limiting the background noise, and increasing the signal-to-noise ratio without modifying the binding of the antibody (Corning Inc., 2001; Xiao and Isaacs, 2012).

3) Addition of the Primary Antibody

In this step, a solution containing the antibody is added to the wells, allowing the antigen to bind to it. This specific antibody that reacts with the protein of interest is referred to as the primary antibody. Unbound antibodies are then washed off the plate using a buffer, typically phosphate-buffered saline or a non-ionic detergent. During each step, the wells are washed at least two times depending on the protocol being followed (Gan and Patel, 2013).

4) Detection

The presence of the primary antibody that is bound is then detected using a secondary, enzyme-coupled antibody. There are various enzymes that can be used in ELISA, including horseradish peroxidase (HRP), alkaline phosphatase (ALP), β-galactosidase, and glucose oxidase. After washing off the excess secondary antibody, a chromogenic substrate for the enzyme is added to yield a visible color change or fluorescence, which indicates the presence of the antigen. Some examples of enzymes, their respective substrates, and the color change associated with a positive reaction is summarized in Table 5.

During the enzyme-substrate reaction, the rate and specificity of the immunological response rely on the catabolic effects of the enzyme (Engvall, 2010). The reaction usually takes 30-60 minutes to complete and can be terminated by adding a suitable solution, such as NaOH, HCl, H_2SO_4, Na_2CO_3, and NaN_3 (Hornbeck, 1992; Crowther, 2009).

5) Final Reading

Depending on the characteristics of the enzyme conjugate and substrate used, results can be read using a spectrophotometer or a fluorometer. Although measurements using a spectrophotometer at 400-600 nm are commonly employed, fluorescent readings have increased sensitivity and are more accurate in determining antigen levels in the sample (Gan and Patel, 2013; Aydin, 2015; Sakamoto, et al., 2017).

Generally, ELISA protocols include placing serial dilutions of the sample in the wells of the microtiter plate. Using these data, a standard curve is constructed, wherein the absorbance reading (y-axis) is plotted against the logarithm scale of the analyte concentration (x-axis). This generates a linear graph (Engvall, 2010; Shah and Maghsoudlou, 2016; Konstantinou, 2017).

Through the years, ELISA became a popular technique in various fields due to its efficiency, ease of use, and speed. It is a highly sensitive and specific test that produces reproducible and quantitative results based on the antibody-antigen coupling approach. Compared to other immunoassays, it is generally safe and cost-effective due to the absence of radioactive substances, large amounts of organic solvents, and sample pre-treatment. Moreover, it requires low-cost and widely available reagents, exhibits robust performance, and is highly accessible in many laboratories (Aydin, 2015; Sakamoto, et al., 2017; Hosseini et al., 2018).

However, ELISA also exhibits some drawbacks. The preparation of the antibody to be used in the study may require sophisticated techniques and expensive culture media. Aside from this, refrigerated transport and storage is necessary to prevent its degradation. ELISA has limited multiplexing options and needs a centralized laboratory equipment. In addition, it requires a relatively high sample volume and may produce false positive/negative results if the blocking of the microtiter plate surface is insufficient. It also exhibits low reliability as compared to RIA (Aydin, 2015; Sakamoto, et al., 2017; Hosseini et al., 2018).

Table 5. Summary of enzyme-substrate reactions and their expected positive results

Enzyme	Substrate	Positive Reaction	Reference
Horseradish peroxidase	*o*-phenylenediamine	Brown color	Aydin, 2015; Ellens and Gielkens, 1980
	5-aminosalicylic acid		
	2,2-diazo-bis(3-ethylbenzothiazoline-6-sulfonic acid)	Blue-green color	Matsuda et al., 1984
	Tetramethylbenzidine	Blue color	Li, et al., 2019
	Luminol	Chemiluminescence	Sakamoto, et al., 2017
Alkaline phosphatase	*p*-nitrophenyl phosphate	Yellow color	Aydin, 2015; Sakamoto, et al., 2017
	Chloro-5-substituted adamantyl-1,2-dioxetane phosphate	Chemiluminescence	Sakamoto, et al., 2017
β-galactosidase	4-methylumbelliferyl galactoside	Fluorescence	Aydin, 2015; Sakamoto, et al., 2017
	Nitrophenyl galactoside	Fluorescence	
Glucose oxidase	Glucose with bromocresol purple	Purple color	Xiong, et al., 2017
	Glucose with mercaptopropionic acid-modified CdTe quantum dots	Fluorescence quenching	Wu, et al., 2018

ELISA Formats Utilized in Functional Food Development

Depending on the target analyte and the mode of immobilization and detection thereof, ELISA can be categorized into the following formats:

direct, indirect, sandwich, and competitive (Figure 3). In functional food development, these formats employ certain modifications in the protocol and vary in the type of detected molecule (antigen/antibody), sensitivity, and specificity. Their similarities and differences are discussed in this section.

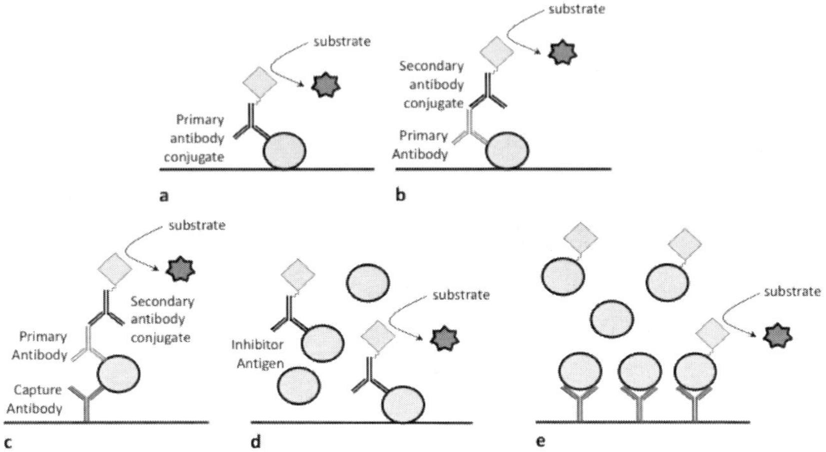

Figure 3. Most common ELISA formats: (a) direct ELISA, (b) indirect ELISA, (c) sandwich ELISA, and competitive ELISA with (d) labeled antibody and (e) labeled antigen (Konstantinou, 2017).

Direct ELISA

First described by Engvall and Perlmann, and Van Weemen and Schuurs in 1971, direct ELISA is the basis for the other ELISA formats. In this method, the antigen of interest is immobilized in the surface of the microtiter plate through passive adsorption. A blocking buffer (e.g., albumin, gelatin, casein, and skimmed milk) is added to saturate all unbound sites and avoid nonspecific interactions. Afterwards, the enzyme-conjugated antibody, which may be monoclonal or polyclonal, is incubated with the immobilized targets, followed by washing to remove unbound antibodies. An appropriate substrate is then added to produce a signal (e.g., color change or emission of light), indicating the presence of the antigen. Due to its direct relationship, the signal increases with increasing antigen concentration (Aydin, 2015; Lin, 2015; Sakamoto, et al., 2017).

Table 6. Direct ELISA employed in the development of functional food

Reference	Functional Food	Antigen/ Antibody Used	Purpose of the Study	Significant Results
Zhang, et al., 2021	Mixture of different natural food ingredients	Serum albumin Inflammatory cytokines	Investigation of the effect of functional food on azoxymethane/ dextran sulfate sodium-induced colorectal cancer	↑ proportions of total T cells, $CD3^+CD4^+$, $CD3^+CD8^+$, IFN-γ-producing $CD4^+$ T cells ↓ IL-6, TNF-α ↑ *Bacteroides, Parasutterella* ↓ *Desulfovibrio*, unclassified Ruminococcaceae
Nouri, Ahari, and Shahbazzadeh, 2018	60 raw cow milk samples (15 with enterotoxin and 45 without)	Antibodies against *Staphylococcus aureus* enterotoxin A	Designing an ELISA kit for detecting *Staphylococcus aureus* enterotoxin A	Detected the toxin in 23% of the samples, with a correctness coefficient of 0.98; with good matching and repetition results, and no cross-reactions; with high sensitivity
Sahasrabudhe, et al., 2018	Commercially extracted lemon pectins with varying degree of methyl esterification (DM)	Isolated TLR2 proteins with the C-terminal of the ligand binding site fused to an HA tag	Elucidation of a new anti-inflammatory mechanism of dietary carbohydrate fibers	Found that pectin inhibits TLR2 and that lower DM pectins can bind more to TLR2 than high-DM pectins

Table 6. (Continued)

Reference	Functional Food	Antigen/ Antibody Used	Purpose of the Study	Significant Results
Chunglok, et al., 2011	Commercial UHT milk, fat-free (0% fat) and plain milk (9% fat)	Rabbit anti *Salmonella* sp. antibody with bioconjugated HRP and single-walled carbon nanotubes	Detection of pathogenic *Salmonella enterica* serovar *Typhimurium*	LOD = 10^3 CFU/mL Showed detection limit 100x greater than commercial antibody/HRP conjugate
Koenen, et al., 2004	Strains of lactic acid bacteria	Serum antibodies to 2,4,6-trinitrophenyl (TNP); mouse monoclonal antibodies CVI-ChIgM59.7 and CVI-ChIgG-47.3	Development of a new selection tool for evaluating the immunomodulating properties of lactic acid bacteria in chicken	Developed an *in vitro* assay based on enhancing the proliferation of sub- optimally stimulated lymphocytes by probiotic bacteria; used only 10 chickens to test 10-15 strains in less than a week; showed no false negative results

Since only one antibody is used, direct ELISA is the simplest and the most time-efficient of all ELISA formats. Cross-reactivity with a secondary antibody will also be prevented in this manner. However, due to the same reason, direct ELISA has lower sensitivity as compared to other formats. Aside from this, labelling the antibody with an enzyme for detection may also lead to its inactivation (Lin, 2015; Sakamoto, et al., 2017).

Direct ELISA is suitable for the qualitative and quantitative analysis of macromolecules and other high molecular weight antigens. Specifically, it is used for antigen detection in a sample, antibody screening, and epitope mapping (Hornbeck, 1992; Lin, 2015). In functional food development, direct ELISA is commonly performed to detect food allergens and other food contaminants, as well as to investigate the molecular mechanisms of the added substances inducing health benefits (Table 6).

Indirect ELISA

Indirect ELISA is an ELISA format wherein the target protein is adsorbed to the wells of a microtiter plate through passive adsorption. Unlike direct ELISA, signal generation involves a two-step process – 1) binding of an unlabeled primary antibody to the immobilized target protein and 2) binding of an anti-species enzyme conjugated secondary antibody to the primary antibody (Lin, 2015; Shah and Maghsoudlou, 2016; Alhajj and Farhana, 2021).

The inclusion of the latter step entails several advantages over direct ELISA. The assay sensitivity of the indirect format is increased due to signal amplification brought about by more than one labeled secondary antibody binding to several epitopes of the primary antibody. Since the primary antibody is unlabeled, maximum immunoreactivity is also ensured because no interference would hinder the activity of its binding sites. Compared to direct ELISA, the indirect format is observed to be more economical because it requires fewer labeled antibodies – one type of secondary antibody can be used to label all primary antibodies. Since there is already a wide variety of secondary antibodies available commercially, the cost associated with individual labeling of primary antibodies is decreased. Indirect ELISA also offers greater flexibility due to the possibility of using different primary antibodies from the same species (Bio-Rad, 2021; Berthold Technologies GmbH and Co KG, 2021; Boster Bio, 2021).

However, these advantages involve a more complex workflow due to the additional incubation step for the introduction of the secondary antibody into the system. This additional step also poses the concomitant risk of cross-reactivity between the primary and secondary antibodies, which leads to the generation of nonspecific signals and, consequently, higher background noise that could potentially decrease the overall net signal. Thus, optimization of the employed protocol is needed to prevent this concern (Bio-Rad, 2021; Berthold Technologies GmbH and Co KG, 2021; Boster Bio, 2021).

Table 7. Research studies employing indirect ELISA for the development of functional foods

Reference	Functional Food	Antigen/ Antibody Used	Purpose of the Study	Significant Results
Haghighi, et al., 2005	Probiotic drink containing *Lactobacillus acidophilus, Bifidobacterium bifidum*, and *Streptococcus faecalis*	Bovine serum albumin and tetanus toxoid	Examination of the effects of the probiotic drink on the induction of chicken antibody response to various antigens	Antibody response to SRBC: Probiotic-treated > nontreated Antibody response to BSA: probiotic-treated = nontreated Antibody response to TT: probiotic-treated = nontreated
Kiyota, et al., 2017	Fresh tomato and tomato-containing processed foods	Rabbit polyclonal anti-Sola I 1 peptide antibodies	Detection of the tomato allergen profilin (Sola I 1)	LOD = 31.3 ng/mL (CV <30%) Quantification range = 62.5 - 500 ng/mL (CV <10%)
Foltz, et al., 2008	Soy, pea, potato, casein, and whey protein hydrolysates	Antibody against cholecystokinin (CCK)-8S; streptavidin-horseradish peroxidase with TMB substrate for detection	Investigation of the CCK-releasing properties of different protein hydrolysates	Induced CCK release at >0.1 mg/L; no significant differences in CCK-releasing properties among protein hydrolysates based on ELISA
Patrick, Hans, and Angelika, 2009	Wines	Polyclonal anti-bovine casein antibodies	Characterization and detection of allergic bovine casein and caseinate fining agents in wines	For caseinate: LODC = 0.5 ppb LOD = 5 ppb intraassay variation = 0.4 – 11.5% interassay variation = 5.8 – 13.6% For α-Lactalbumin: LODC = 2.4 mg/L LOD = 5.2 mg/L For β-Lactoglobulin: LODC = 0.8 mg/L LOD = 1.8 mg/L

Reference	Functional Food	Antigen/ Antibody Used	Purpose of the Study	Significant Results
Schneid, et al., 2006	Chicken meat	Monoclonal antibody against the outer membrane protein of *Salmonella enterica* serovar *Enteritidis*	Detection of *Salmonella* in 154 samples of chicken meat	Accuracy = 94% sensitivity = 94% specificity = 94% PPV = 82% NPV = 98%

In functional food development, ELISA formats are usually implicated in the efficacy and safety step of the so-called cycle of innovation. Similar to sandwich ELISA, indirect ELISA is preferably used in authenticity testing of food products, specifically in the detection and identification of allergen ingredients (Asensio et al., 2008). Aside from assessing food safety, indirect ELISA is also employed in the efficacy testing of functional foods, particularly in *in vivo* studies wherein the determination of total antibody concentration in samples is required (Bio-Rad, 2021). Examples of these studies are enumerated in Table 7.

Sandwich ELISA

Developed by Kato and his colleagues in 1977, the unique feature of sandwich ELISA is its use of a capture antibody, which is first added to the microtiter plate to coat its surface. After blocking, the antigen-containing sample is added and allowed to react with the immobilized capture antibody. The bound antigen is then "sandwiched" with an enzyme-linked antibody for quantification after substrate addition. An alternative to this direct technique is an indirect method wherein a primary antibody is applied, followed by the addition of an enzyme-labeled secondary antibody for detection (Gan and Patel, 2013; Aydin, 2015; Sakamoto, et al., 2017).

Since at least two antibodies are used, sandwich ELISA exhibits greater sensitivity and specificity (i.e., two to five times greater) as compared to other ELISA formats. The use of a capture antibody also eliminates the purification step for the antigen, which allows the evaluation of more complex substances. However, the preparation of two antibodies is expensive and labor-intensive. Adding a capture antibody also increases the assay time and requires more optimization steps to avoid cross-reactivity between the antibodies used (Gan and Patel, 2013; Aydin, 2015; Sakamoto, et al., 2017).

Due to its increased sensitivity and specificity, sandwich ELISA is more commonly performed than direct and indirect formats in detecting food allergens and quantifying essential proteins and macromolecules for functional food development. Examples of such studies are detailed in Table 8.

Table 8. Sandwich ELISA used in functional food development

Reference	Functional Food	Antigen/ Antibody Used	Purpose of the Study	Significant Results
Moriyama, et al., 2020	Processed soybean foods (soybean flour, soymilk, Kori-tofu, natto)	Antibodies against β-conglycinin	Quantification of β-conglycinin in processed soybean foods using pretreatment and commercially available ELISA kits	Pretreatment by mixing the samples with rice flour prior to protein extraction; accurately quantified β-conglycinin in soybean samples with good reproducibility (correction factor of 0.24)
Gujral, et al., 2017	Therapeutic and functional products, chicken egg yolks	Anti-phosvitin polyclonal IgY (with and without biotinylation)	Development of a double antibody sandwich ELISA (DAS-ELISA) method for quantifying phosvitin content	Detection range: DAS-ELISA = 16.8–90 ng/mL Biotinylated DAS-ELISA = 7.5–40 ng/mL Biotinylated DAS-ELISA showed higher accuracy and sensitivity for quantifying phosvitin

Reference	Functional Food	Antigen/ Antibody Used	Purpose of the Study	Significant Results
Lacorn, et al., 2016	Raw and processed food samples	Soy-specific monoclonal antibodies conjugated to peroxidase	Quantification of soy proteins (glycinin) from food samples collected from the market	No cross-reactivities; High recoveries even from highly processed samples; high precision; suitable for detecting unheated to strongly heated soy proteins in various foods
Xu, et al., 2016	Local milk samples and soybean powder (heat-treated and unheated)	HRP-labeled monoclonal antibodies against Kunitz trypsin inhibitor (KTI)	Development of sandwich ELISA and immunochromato-graphic strip (ICS) method for the semi-quantification of KTI	LOD = 0.1481 ng/mL Linear dynamic range = 0.2–20 ng/mL Recovery = 98.40–110.66%, with CV = 4.41–10.62% Negligible cross-reactivities
Venkatara-mana, et al., 2015	195 field samples of maize, wheat, and rice	Polyclonal antisera and monoclonal antibodies against Ochratoxin A	Development of sandwich dot-ELISA for specific detection of Ochratoxin A	LOD = 5.0 ng/mL No cross-reactivity with other tested mycotoxins; better alternative for routine cereal-based food and feed analysis using conventional HPLC
Chávez, et al., 2012	Raw milk, yogurt, dietary supplement, milk-like probiotic and soya drink	Glycomacro-peptide polyclonal antiserum	Determination of glycomacro-peptide for detecting liquid whey in raw milk	Sandwich ELISA has an appropriate specificity and better limit of detection to analyze raw milk for presence of cheese whey compared with colorimetric method, SDS-PAGE, immunoblot, and immunochromato-graphic test.

Table 8. (Continued)

Reference	Functional Food	Antigen/ Antibody Used	Purpose of the Study	Significant Results
Ecker, and Cichna-Markl, 2012	Local food samples containing lupine (tofu, rusk, shortbread, crisp bread, bread, patties, spread, noodles)	Chicken anti-lupine IgY (coating antibody), rabbit anti-lupine IgG (secondary antibody)	Development and validation of a sandwich ELISA for detecting traces of lupine proteins in foods	Successful detection of proteins from white (*Lupinus albus*), blue, (*Lupinus angustifolius*), and yellow (*Lupinus luteus*) lupine; no cross-reactivity with 34 plant species potentially present in the food samples; detection of lupine flour in both unprocessed and processed doughs at 1 ppm
Pelaez-Lorenzo, et al., 2010	Milk; Meat, fish, or cereal and fruit baby foods with and without dairy products (milk, cheese, or yogurt)	Polyclonal antiβ-lactoglobulin (antiβLG)	Optimization of sandwich ELISA method for detecting β-lactoglobulin in baby foods	Optimized ELISA format allows detection of standard βLG at the picomolar level; can detect 1 part of dairy product in 1000 parts of baby food
Lee and Kim, 2010	Food samples with egg ingredients (raw hen's egg, hard-boiled quail's egg, egg biscuit, mayonnaise, dried pollack soup, and ice cream) and without egg ingredients (milk, orange juice, biscuit, chicken meat, and duck roast)	Ovomucoid-mouse monoclonal antibody, egg protein mixture-rabbit polyclonal antibody, ovalbumin-rabbit polyclonal antibody, and lysozyme-rabbit polyclonal antibody	Detection of hidden egg allergens (egg whole protein, egg white protein, egg yolk protein, ovomucoid, ovalbumin, ovotransferrin, lysozyme, and α-livetin) in various foods	High sensitivity and specificity of sandwich ELISA (initial screening and immunoblotting (final test) in detecting potentially allergenic egg residues in processed foods in trace amounts; using ELISA alone showed cross-reactivity with matrix components, leading to false-positive results

Reference	Functional Food	Antigen/ Antibody Used	Purpose of the Study	Significant Results
Fæste, and Plassen, 2008	Fish and food samples (white sauce, fish soup, soy sauce, mushroom soup, bread, wine)	Polyclonal rabbit anti-cod parvalbumin antibody (capture), biotinylated conjugate of same antibody (detection)	Quantitation of fish in food matrices using muscle protein parvalbumin	High specificity for fish and no cross-reactivity with other species Recovery = 68–138% Precision = <12% (intra-assay) and <19% (inter-assay) LOD = 0.01 mg parvalbumin/kg food or ~5 mg fish/kg food Can detect fish protein traces at sufficiently low levels to prevent allergic reactions

Competitive ELISA

Competitive ELISA, otherwise known as inhibition ELISA or blocking ELISA, is an ELISA format wherein the inhibitor antigen is pre-coated to the wells of a microtiter plate. Unlike other formats, competitive ELISA, as the name implies, uses a competitive reaction process between the sample antigen and the inhibitor antigen, and the primary antibody (Bio-Rad, 2021; Boster Bio, 2021).

Considered to be the most complex format, signal generation in competitive ELISA is achieved in three steps. The first one involves the incubation of the primary antibody with the crude sample, allowing the binding and formation of complexes between the primary antibody and the sample antigen, if there is any. In this step, the amount of antibody is greater than the amount of antigen and, thus, there are unbound primary antibodies left. Second, the resulting antigen-antibody mixture is added to the pre-coated wells wherein the remaining unbound primary antibodies can bind to the adsorbed inhibitor antigen, while those bound to the sample antigen are washed off. For the third step, enzyme conjugated secondary antibodies are added to bind to the primary antibody-inhibitor antigen complexes and produce signals. In this ELISA format, the amount of sample antigen is inversely proportional to the signal intensity – the higher the concentration of sample antigen, the lower the signal generated (Shah and Maghsoudlou, 2016; Lu, 2017; Bordenave and Ferruzzi, 2018; Alhajj and Farhana, 2021; Berthold Technologies GmbH and Co KG, 2021; Cusabio Technology LLC, 2021).

As previously mentioned, crude or impure samples can be used in competitive ELISA. This is due to its observed high assay sensitivity even in complex environments, an advantage that makes it the most suitable format for the identification and quantification of small quantities of sample antigen. Furthermore, it generates more consistent results as evidenced by less variability in signals produced between duplicate samples and assays. It also offers maximum flexibility because it can be based on other formats, such as direct, indirect, or sandwich ELISA (Lu, 2017; Bordenave and Ferruzzi, 2018; Bio-Rad, 2021; Boster Bio, 2021; Berthold Technologies GmbH and Co KG, 2021; Cusabio Technology LLC, 2021).

Similar to indirect and sandwich ELISA, competitive ELISA is also implicated in the efficacy and safety step in the cycle of innovation used in functional food development, specifically in the identification and quantification of allergen ingredients. Findings generated from these studies support the applicability and reliability of indirect competitive ELISA for control strategies in the food industry and in allergy prevention. In addition to its potential role in authenticity testing of food products, the competitive ELISA format can also be used in the detection and quantification of viable probiotic cells in processed products, such as fermented milk and yogurt. Through this, it is presumed that the relationship between the composition of microflora present in food products and their observed beneficial effects will be better understood (Garrote, et al., 2005). Research studies involving competitive ELISA for functional food development are detailed in Table 9.

Table 9. Competitive ELISA involved studies focusing on functional food development

Reference	Functional Food	Antigen/Antibody Used	Purpose of the Study	Significant Results
Holzhauser and Vieths, 1999	Complex food matrices (e.g., legumes, nuts, and stone fruits, etc.)	Monoclonal antibody against Ara h 1	Detection of hidden peanut protein residues in various foods	LOD = 2 ppm LOQ = 4 ppm
Castillo and Cassola, 2017	Raw and processed foodstuffs (e.g., almond, hazelnut, barley, etc.)	Monoclonal antibodies against β-casein N and D	Detection and quantification of traces of β-casein in raw and processed foodstuffs	For 1H3-based icELISA: LOD = 0.29 ppm LOQ = 0.80 ppm intraassay CV = 0.47 - 9.10% interassay CV = 0.37 - 6.33%
Castillo and Cassola, 2017	Raw and processed foodstuffs (e.g., almond, hazelnut, barley, etc.)	Monoclonal antibodies against β-casein N and D	Detection and quantification of traces of β-casein in raw and processed foodstuffs	For 6A12-based icELISA: LOD = 0.98 ppm LOQ = 2.62 ppm intraassay CV = 0.41 - 4.71% interassay CV = 1.38 - 5.93%
Singh, et al., 2019	Grain-based foods (e.g., rye bread, etc.)	Polyclonal antibody against citrinin	Detection and quantification of citrinin in grain-based food samples	LOD = 0.04 ± 0.007 ng/mL LOQ = 0.10 ± 0.02 ng/mL No cross-reactivity with other mycotoxins; icELISA method comparable to the instrumental methods used by other laboratories in the PT testing scheme

Table 9. (Continued)

Reference	Functional Food	Antigen/ Antibody Used	Purpose of the Study	Significant Results
Xu, et al., 2019	Commercially available hypoglycemic functional food samples	Anti-rosiglitazone (RSG) monoclonal antibody (4G6)	Development of an immuno-chromatographic strip for detection of RSG in functional foods	IC_{50} = 1.0889 ng/mL LOD = 0.2178 ng/mL Linear detection range = 0.2178–1.742 ng/mL
Burkin and Galvidis, 2010	Milk, eggs, and honey	Polyclonal antibody against lincomycin	Determination of lincomycin in food matrices such as milk, eggs, and honey	LOD (milk) = 0.43 ng/mL LOD (eggs) = 0.65 ng/mL LOD (honey) = 1.9 ng/mL Cross-reactivity of clindamycin = 111; test may be considered as a group-specific assay for lincosamides
Vass, et al., 2008	Incurred porcine muscle samples, baby foods	Polyclonal antibodies against 3-carboxyphenyl semicarbazide (rabbit antisera)	Development of ELISA formats (indirect and direct) for detecting trace amounts of low-molecular weight semicarbazide	Linear readout = 0.01-0.2 μg/L IC_{50} = 0.06 - 2.28 μg/L Recovery level = 82.9 to 105.3%, CV > 15.5% Detection capability and threshold = 0.3 and 0.11 μg/kg

Throughout the years, ELISA has been used in the analysis and quantification of various antigens and antibodies. As discussed in the previous paragraphs, each ELISA format is suitable for certain purposes and thereby should be used according to its own advantages and disadvantages, as summarized in Table 10.

Table 10. Overall comparison among ELISA formats

Format	Screening	Advantages	Disadvantages
Direct	Antibody	Simple and rapid, Less susceptible to errors	Low sensitivity, Possible inactivation of antibody
Indirect	Antibody	High sensitivity and versatility	Nonspecific signal due to cross-reactivity of secondary antibody
Sandwich	Antigen	Very high sensitivity and specificity, Suitable for complex samples	Labor-intensive and expensive, Requires more optimization steps due to cross-reactivity
Competitive	Antigen / Antibody	High sensitivity, specificity, and versatility Very robust – minimizes sample dilution and matrix effects	Time-consuming Very complex protocol

Other Applications of ELISA in Functional Food

Aside from its role in functional food development, the previously described ELISA formats are also implicated in functional food analysis and monitoring, especially in ensuring the safety, quality, and authenticity of functional foods produced and released in the market. This includes detection of food contaminants, such as pesticides, antibiotics, and microbes. In the study conducted by Abad, et al., (2001), an indirect competitive ELISA was developed for the routine monitoring of thiabendazole residues in fruit juices. Assay sensitivity was observed to be significantly lower than the allowable tolerances of thiabendazole and its metabolite in or on food commodities – 1 ng/mL in orange and grapefruit juices, 5 ng/mL in banana juice, and 20 ng/mL in apple and pear juices – thereby supporting its suitability as a method of analysis for thiabendazole monitoring (Abad, et al., 2001; National Center for Biotechnology Information, 2021). Similarly, in the study conducted by Galarini, et al., (2014), the indirect competitive ELISA used to screen sulfonamide residues in the muscle, eggs, milk, honey, and feed exhibited detection capabilities that were significantly lower than the lethal dose of sulfonamides – lower or equal to 20 μg/kg for muscle, eggs and milk, 10 μg/kg for honey, and 2 mg/kg for feed (National Center for Biotechnology Information, 2021). In the study conducted by Zhu et al., (2016), on the other hand, a sandwich ELISA is used for the rapid detection of *Bacillus cereus* in food as a means to prevent food contamination and food-borne outbreaks. Linear detection range of the method was determined to be approximately $1 \times 10^4 - 2.8 \times 10^6$ cells/mL with a limit of detection of 0.9×10^3 cells/mL, a concentration that is also significantly lower than the minimum concentration of *B. cereus* that is known to cause illness (Zhu, et al., 2016; Schneider, et al., 2021).

Future Directions: Functional Food Monitoring and Regulations for Better Health and Wellness

Development of Functional Food in the Philippines - Current and Future Approaches

The concept of functional food is not new in the Philippine context. Traditional Filipino foods, such as rice, coconut, legumes, native vegetables, and fruits, are known to provide benefits beyond their nutritive value (Tecson-Mendoza, 2007). Aside from natural sources of foods, Philippine fermented foods are part of the Filipino culture and are often associated with different ethnic groups (Banaay, et al., 2013). The principles and technology of Philippine fermented foods are extensively discussed by Sanchez (2008). Research initiatives on exploring the safety and nutritional quality of traditional fermented food products found promising LAB strains that can be developed as a functional food. Recent studies in the biomedical, industrial, and food applications of local LAB strains isolated from fermented foods succeeding the data presented by

Banaay, et al., (2013) are discussed in this section. LAB isolates from muscovado-based kefir from Leyte, Philippines showed antibacterial activities against pathogens (Elegado, et al., 2016). *Lactobacillus fermentum* 4B1, another probiotic candidate isolated from *balao-balao*, was observed to reduce obesity in mice (Balolong, et al., 2017). Meanwhile, bacteriocinogenic strains, *L. plantarum* BS25 (LPBS25) and *Pediococcus acidilactici* 3G3 (PG3G3), which were isolated from *balao-balao* (fermented rice-shrimp mixture), showed cytotoxic activity against colorectal cancer cells (Asi, et al., 2020). These studies highlight the importance of tapping local LAB strains as starter cultures for the development of functional foods in the Philippines.

Unfortunately, reports on the development and regulation of Philippine functional foods and their potential as adjunct therapy are still lacking. A preliminary report demonstrated that LPBS25-fermented dark chocolate drink reduced the cholesterol levels *in vitro* and in hypercholesterolemic mice (Tumampo, et al., 2016). Inoculation of *L. plantarum* BS and PG3G3 in *pindang damulag* (fermented carabeef) improved the physicochemical, microbiological, and sensory properties of the fermented food (Balmori, et al., 2019). Likewise, functional foods with prebiotic properties are still being explored and developed as of writing (Luci-Atienza, 2021). With this, ELISA can serve as a promising approach in addressing such deficiencies and improving Philippine functional foods in the future.

Using ELISA for the Development of New Functional Foods

A previous study highlighted the use of ELISA for the development of new and efficient functional foods from insulin-like antigens in sprouts (Pathak and Martirosyan, 2011). Likewise, ELISA has been used to monitor and detect adulteration and contamination in commercially available functional foods. For instance, an ELISA method was optimized to detect β-lactoglobulin in baby foods (Pelaez-Lorenzo, et al., 2010). Synthetic drugs are often added in marketed functional foods to attract consumers on the claimed therapeutic effect of these drugs, which are commonly falsified as natural substances (Song, et al., 2012). A solid-phase extraction ELISA was developed to monitor the presence of 17β-19-nortestosterone, a steroid that may cause several metabolic complications, in anti-fatigue functional foods (Zhang, et al., 2009). A direct competitive ELISA was optimized to screen out sildenafil in functional foods (Song, et al., 2012). Pecan residues in pecan nuts, a popular processed product due to their health benefits, were detected using a sandwich-type ELISA (Tran, 2013). A highly sensitive sandwich ELISA was developed by Chávez, et al., (2012) to detect fraudulent addition of cheese whey to milk by screening for the presence of glycomacropeptide (GMP). Although the presence of GMP in milk is not a health hazard, it affects the quality and downstream processing of dairy products. Interestingly, GMP can be tapped as an additive ingredient to produce functional foods, thus, this ELISA system can also be used to develop GMP-fortified functional foods (Chávez, et al., 2012). These studies demonstrate the good accuracy and reproducibility of ELISA as a quantitative and qualitative test for the development and monitoring of new functional foods. Furthermore, ELISA presents a practical and user-friendly diagnostic approach to industries and research institutes. In the Philippines, ELISA is widely used to develop assay-based diagnosis for under-recognized public health problems, including tests for leptospirosis (Kitashoji, et al., 2015), dengue (Lopez, et al., 2021), and schistosomiasis (Angeles, et al., 2020). This shows that ELISA is being developed as a practical diagnostic and validation test and, thus, studies utilizing ELISA for the development of Philippine functional foods should be prioritized in the future.

CONCLUSION

The popularity of functional foods is continually rising among consumers because not only do they promote optimal health and prevent chronic diseases, but they consequently reduce healthcare costs and improve the consumers' quality of life as well. With this, extensive research is being conducted to explore and further understand functional foods, beverages, and ingredients. These studies usually include the use of biomarkers to determine the effects of the administered functional foods to the organism. However, reported biomarkers are commonly associated with monitoring disease states and progression, rather than evaluating the health status of an individual. Since the main goal of functional foods is to boost optimal health, it is essential to identify biomarkers at the very early stages, even before the onset of the disease. This would serve as health and/or prevention markers instead of disease markers. Moreover, nutritional biomarkers should also be considered in assessing the overall health status of an individual. Nevertheless, clinical biomarkers are highly relevant in the development and regulation of functional foods.

Despite the wide variety of Philippine traditional and fermented foods of pharmaceutical and nutritional importance, the development of functional foods in the country is still insufficient. Most research studies published mainly focus on the potential of local LAB strains in biomedical, industrial, and food applications. Undoubtedly, ELISA could be applied in functional food development and regulation in the Philippines. As discussed in this chapter, due to the immunological concept of antigen-antibody interactions in ELISA, detection and quantification of certain substances can be performed with ease. Still, this technique has mostly been applied to address under-recognized public health issues in the country. Hence, the use of ELISA in developing and regulating functional foods from local food sources that can serve as an adjunct therapy against chronic diseases could pose a promising area of study for future research investigations.

REFERENCES

Abad, Antonio, Juan J. Manclús, María J. Moreno, and Angel Montoya. 2001. "Determination of thiabendazole in fruit juices by a new monoclonal enzyme immunoassay." *Journal of AOAC International, 84(1)* 156-161.

Adair, John C., Julius Charlie, John E. Dencoff, Jeffrey A. Kaye, Joseph F. Quinn, Richard M. Camicioli, William G. Stetler-Stevenson, and Gary A. Rosenberg. 2004. "Measurement of gelatinase B (MMP-9) in the cerebrospinal fluid of patients with vascular dementia and Alzheimer disease." *Stroke 35:e159–e162.* Accessed July 27, 2021. PubMed: 15105518.

Aguilar-Toalá, J. E., R. Garcia-Varela, H. S. Garcia, V. Mata-Haro, A. F. González-Córdova, B. Vallejo-Cordoba, and A. Hernández-Mendoza. 2018. "Postbiotics: An evolving term within the functional foods field." *Trends in Food Science and Technology, 75* 105–114. Accessed July 27, 2021. https://doi.org/10.1016/j.tifs.2018.03.009.

Albert, Marilyn S., Steven T. DeKosky, Dennis Dickson, Bruno Dubois, Howard H. Feldman, Nick C. Fox, Anthony Gamst, et al. 2011. "The diagnosis of mild cognitive impairment due to Alzheimer's disease: recommendations from the National Institute on Aging–Alzheimer's Association workgroups on diagnostic guidelines for Alzheimer's disease." *Alzheimer's & Dementia, 7* 270–279.

Alhajj, Mandy, and Aisha Farhana. 2021. *Enzyme Linked Immunosorbent Assay.* February 6. Accessed July 27, 2021. https://www.ncbi.nlm.nih.gov/books/NBK555922/.

Al-Sheraji, Sadeq Hasan, Amin Ismail, Mohd Yazid Manap, Shuhaimi Mustafa, Rokiah Mohd Yusof, and Fouad Abdulrahman Hassan. 2013. "Prebiotics as functional foods: A review." *Journal of Functional Foods, 5(4)* 1542-1553. Accessed July 27, 2021. https://doi.org/10.1016/j.jff.2013.08.009.

American Psychiatric Association. 2013. *Diagnostic and Statistical Manual of Mental Disorders 5th ed.* Washington, D.C.: American Psychiatric Association.

Amoranto, Mia Beatriz C., and Marilen Parungao Balolong. 2020. "Immunomodulation in Lactic Acid Bacteria: Exploring Prospects for Adjunct Functional Food Therapy." *Philippine Journal of Health Research and Development, 24(3)* 64-76.

Andreasen, Niels, Lennart Minthon, Pia Davidsson, Eugeen Vanmechelen, Hugo Vanderstichele, Bengt Winblad, and Kaj Blennow. 2001. "Evaluation of CSF-tau and CSF-Abeta42 as diagnostic markers for Alzheimer disease in clinical practice." *Arch Neurol., 58(3)* 373-379. Accessed July 27, 2021. doi:10.1001/archneur.58.3.373.

Angeles, Jose Ma. M., Yasuyuki Goto, Masashi Kirinoki, Elena A. Villacorte, Kharleezelle J. Moendeg, Pilarita T. Rivera, Yuichi Chigusa, and Shin-ichiro Kawazu. 2020. "Field evaluation of recombinant antigen ELISA in detecting zoonotic schistosome infection among water buffaloes in endemic municipalities in the Philippines." *Frontiers in Veterinary Science, 7: 592783.*

Angeles, Jorge Gil C., Jeric C. Villanueva, Lawrence Yves C. Uy, Sheila Mae Q. Mercado, Maria Claret L. Tsuchiya, Jickerson P. Lado, Mark Rickard N. Angelia, Mia Clare Marie Bercansil-Clemencia, Maria Amelita C. Estacio, and Mary Ann O. Torio. 2021. "Legumes as Functional Food for Cardiovascular Disease." *Applied Sciences, 11(12): 5475* Accessed July 27, 2021. https://doi.org/10.3390/app11125475.

Asensio, Luis, Isabel González, Teresa García, and Rosario Martín. 2008. "Determination of food authenticity by enzyme-linked immunosorbent assay (ELISA)." *Food Control, 19* 1-8.

Asi, Ranelle Janine L., Christian Luke D. Badua, Jason Bryan D. Atun, Joshua Ismael L. Lapus, Sheena May O. Lopez, Sonia D. Jacinto, Ahmad Reza F. Mazahery, and Marilen P. Balolong. 2020. "An optimized de Man Rogosa Sharpe-McCoy's 5A medium formulation for the in vitro screening of cytotoxic effects by probiotic candidates against colorectal cancer cells." *Philippine Science Letters, 13(2)* 139-148.

Atlante, Anna, Giuseppina Amadoro, Antonella Bobba, and Valentina Latina. 2020. "An Approach to Modulate Molecular Mechanisms of Alzheimer's Disease." *Cells, 9(11): 2347* Accessed July 27, 2021. doi: 10.3390/cells9112347.

Aydin, Suleyman. 2015. "A short history, principles, and types of ELISA, and our laboratory experience with peptide/protein analyses using ELISA." *Peptides, 72* 4-15. Accessed July 27, 2021. https://doi.org/10.1016/j.peptides.2015.04.012.

Badimon, Lina, Esther Peña, Gemma Arderiu, Teresa Padró, Mark Slevin, Gemma Vilahur, and Gemma Chiva-Blanch. 2018. "C-reactive protein in atherothrombosis and angiogenesis." *Frontiers in Immunology, 9:430* Accessed July 27, 2021. https://dx.doi.org/10.3389%2Ffimmu.2018.00430.

Bae, Yoon Ju, Michael Schaab, and Juergen Kratzsch. 2015. "Calcitonin as biomarker for the medullary thyroid carcinoma." *Recent results in cancer research, Fortschritte der Krebsforschung Progres dans les recherches sur le cancer, 204* 117-137. Accessed July 27, 2021. https://doi.org/10.1007/978-3-319-22542-5_5.

Balmori, Vernabelle, Erlinda I. Dizon, A. S. A. Barrion, and Francisco B. Elegado. 2018. "Effect of adjunct inoculation of lactobacillus plantarum BS and pediococcus acidilactici 3G3 on the microbiological, physicochemical and sensory properties of fermented carabeef (*Pindang damulag*)." *Food Research, 3(1)* 70-78. Accessed July 27, 2021. https://doi.org/10.26656/fr.2017.3(1).221.

Balolong, Marilen P., R. L. S. Bautista, N. C. A. Ecarma, E. C. Balolong Jr., A. V. Hallare, and Francis B. Elegado. 2017. "Evaluating the anti-obesity potential of *Lactobacillus fermentum* 4B1, a probiotic strain isolated from balao-balao, a traditional Philippine fermented food." *International Food Research Journal, 24(2)* 819-824.

Banaay, Charina Gracia B., Marilen P. Balolong, and Francisco B. Elegado. 2013. "Lactic Acid Bacteria in Philippine Traditional Fermented Foods." In *Lactic Acid Bacteria - R & D for Food, Health and Livestock Purposes*, by Marcelino Kongo, Accessed July 27, 2021. https://doi.org/10.5772/50582. IntechOpen.

Bang, Yung-Jue, Eric Van Cutsem, Andrea Feyereislova, Hyun C. Chung, Lin Shen, Akira Sawaki, Florian Lordick, et al. 2010. "Trastuzumab in combination with chemotherapy versus chemotherapy alone for treatment of HER2-positive advanced gastric or gastro-oesophageal junction cancer (ToGA): A phase 3, open-label, randomised controlled trial." *Lancet, 376(9742)* 687-697. Accessed July 27, 2021. https://doi.org/10.1016/s0140-6736(10) 61121-x.

Bast Jr., Robert C., Peter Ravdin, Daniel F. Hayes, Susan Bates, Herbert Fritsche Jr., John M. Jessup, Nancy Kemeny, Gershon Y. Locker, Robert G. Mennel, and Mark R. Somerfield. 2001. "Update of recommendations for the use of tumor markers in breast and colorectal cancer: clinical practice guidelines of the American Society of Clinical Oncology." *Journal of Clinical Oncology, 19* 1865-1878.

Berthold Technologies GmbH & Co KG. 2021. *ELISA Formats.* June 20. Accessed June 20, 2021. https://www.berthold.com/en/bioanalytic/applications/elisa/formats/.

Bio-Rad. 2021. *Chapter 2 - ELISA Technology.* June 19. Accessed June 19, 2021. https://www.bio-rad-antibodies.com/static/2017/an-introduction-to-elisa/elisa-basics-guide.pdf?_ga=2.121367688.1047634877.1624101385-1948381072.1624101385.

Biörklund, M., A. van Rees, R. Mensink, and G. Önning. 2005. "Changes in serum lipids and postprandial glucose and insulin concentrations after consumption of beverages with β-glucans from oats or barley: A randomised dose-controlled trial." *European Journal of Clinical Nutrition, 59(11)* 1272-1281. Accessed July 27, 2021. https://doi.org/10.1038/sj.ejcn.1602240 .

Bjartell, Anders S. 2013. "Next-generation prostate-specific antigen test: ready to use?" *European Urology, 64(5)* 700-702. Accessed July 27, 2021. https://doi.org/10.1016/j.eururo.2013.06.052.

Bordenave, Nicolas, and Mario G. Ferruzzi. 2018. *Functional Food and Beverages: In vitro Assessment of Nutritional, Sensory, and Safety Properties.* New Jersey: John Wiley & Sons.

Boster Bio. 2021. *ELISA Fundamental Principle, How It Works.* June 20. Accessed June 20, 2021. https://www.bosterbio.com/protocol-and-troubleshooting/elisa-principle.

Bridel, Claire, Wessel N. van Wieringen, Henrik Zetterberg, Betty M. Tijms, Charlotte E. Teunissen, and NFL Group. 2019. "Diagnostic Value of Cerebrospinal Fluid Neurofilament Light Protein in Neurology: A Systematic Review and Meta-analysis." *JAMA Neurol, 76(9)* 1035-1048. Accessed July 27, 2021. doi:10.1001/jamaneurol. 2019.1534.

Buerger, Katharina, Michael Ewers, Tuula Pirttilä, Raymond Zinkowski, Irina Alafuzoff, Stefan J. Teipel, John DeBernardis, et al. 2006. "CSF phosphorylated tau protein correlates with neocortical neurofibrillary pathology in Alzheimer's disease." *Brain, 129(11)* 3035-3041. Accessed July 27, 2021. https://doi.org/10.1093/brain/awl269.

Burkin, Maksim A., and I. Galvidis. 2010. " Development of a Competitive Indirect ELISA for the Determination of Lincomycin in Milk, Eggs, and Honey." *Journal of Agricultural and Food Chemistry, 58(18)* 9893-9898. Accessed July 27, 2021. http://dx.doi.org/10.1021/jf101731h.

Campmajó, Guillem, Nerea Núñez, and Oscar Núñez. 2019. "The role of liquid chromatography-mass spectrometry in food Integrity and authenticity." In *Mass Spectrometry - Future Perceptions and Applications*, by Ganesh Kamble, Accessed July 27, 2021. doi: 10. 5772/intechopen.78510. IntechOpen.

Casadesus, Gemma, Barbara Shukitt-Hale, Heather M. Stellwagen, Xiongwei Zhu, Hyoung-Gon Lee, Mark A. Smith, and James A. Joseph. 2004. "Modulation of hippocampal plasticity and cognitive behavior by short-term blueberry supplementation in aged rats." *Nutrition Neuroscience, 7(5-6)* 309-316. Accessed July 27, 2021. https://doi.org/10.1080/10284150400020482.

Castillo, Daniela S., and Alejandro Cassola. 2017. "Novel sensitive monoclonal antibody based competitive enzyme-linked immunosorbent assay for the detection of raw and processed bovine beta-casein." *PLOS ONE, 12(7): e0182447* Accessed July 27, 2021. https://doi.org/10.1371/journal.pone.0182447.

Chapman, Paul B., Axel Hauschild, Caroline Robert, John B. Haanen, Paolo Ascierto, James Larkin, Reinhard Dummer, et al. 2011.

"Improved survival with vemurafenib in melanoma with BRAF V600E mutation." *New England Journal of Medicine, 364(26)* 2507-2516. Accessed July 27, 2021. doi: 10.1056/NEJMoa1103782.

Chávez, Norma A., Juan Jauregui, Laura A. Palomares, Karla E. Macías, Mariela Jiménez, and Eva Salinas. 2012. "A highly sensitive sandwich ELISA for the determination of glycomacropeptide to detect liquid whey in raw milk." *Dairy Science and Technology, 92(2)* 121-132. Accessed July 27, 2021. https://doi.org/10.1007/s13594-011-0052-3.

Chen, Lei, David W. Y. Ho, Nikki P. Y. Lee, Stella Sun, Brian Lam, Kwong-Fai Wong, Xin Yi, et al. 2010. "Enhanced detection of early hepatocellular carcinoma by serum SELDI-TOF proteomic signature combined with alpha-fetoprotein marker." *Annals of Surgical Oncology, 17(9)* 2518-2525. Accessed July 27, 2021. https://doi.org/10.1245/s10434-010-1038-8.

Chen, Zhen-Yu, Rui Jiao, and Ka Ying Ma. 2008. "Cholesterol-lowering nutraceuticals and functional foods." *Journal of Agricultural and Food Chemistry, 56(19)* 8761-8773. Accessed July 27, 2021. https://doi.org/10.1021/jf801566r.

Chunglok, Wilanee, Dyah Kinasih Wuragil, Sukunya Oaew, Mithran Somasundrum, and Werasak Surareungchai. 2011. "Immunoassay based on carbon nanotubes-enhanced ELISA for Salmonella enterica serovar Typhimurium." *Biosensors and Bioelectronics, 26(8)* 3584-3589. Accessed July 27, 2021. https://doi.org/10.1016/j.bios.2011.02.005.

Clydesdale, Fergus. 2004. "Functional foods: Opportunities & challenges." *Food Technology Magazine, 58(12)* Accessed July 27, 2021. https://www.ift.org/news-and-publications/food-technology-magazine/issues/2004/december/features/functional-foods-opportunities-and-challenges .

Colleoni, Marco, Giuseppe Viale, David Zahrieh, Giancarlo Pruneri, Oreste Gentilini, Paolo Veronesi, Richard D. Gelber, et al. 2004. "Chemotherapy is more effective in patients with breast cancer not expressing steroid hormone receptors: a study of preoperative treatment." *Clinical Cancer Research, 10(19)* 6622-6628. Accessed July 27, 2021. https://doi.org/10.1158/1078-0432.ccr-04-0380.

Corning Inc., L. S. 2001. *Effective Blocking Procedures - ELISA Technical Bulletin - No. 3.* http://www.labcluster.com/news4_3/Corning_elisa3_4.pdf.

Crowther, John R. 2009. "Stages in ELISA." In *The ELISA Guidebook*, by John R. Crowther, 43-78. New Jersey: Humana Press Inc.

Cusabio Technology LLC. 2021. *Four Types of ELISA*. June 20. Accessed June 20, 2021. https://www.cusabio.com/c-20659.html.

Daiber, Andreas, Ning Xia, Sebastian Steven, Matthias Oelze, Alina Hanf, Swenja Kröller-Schön, Thomas Münzel, and Huige Li. 2019. "New therapeutic implications of endothelial nitric oxide synthase (eNOS) function/dysfunction in cardiovascular disease." *International Journal of Molecular Sciences, 20(1):187* Accessed July 27, 2021. https://doi.org/10.3390/ijms20010187.

Dematteo, Ronald P., Karla V. Ballman, Cristina R. Antonescu, Robert G. Maki, Peter W. T. Pisters, George D. Demetri, Martin E. Blackstein, et al. 2009. "Adjuvant imatinib mesylate after resection of localised, primary gastrointestinal stromal tumour: a randomised, double-blind, placebo-controlled trial." *The Lancet, 373(9669)* 1097-1104. Accessed July 27, 2021. https://doi.org/10.1016/s0140-6736(09)60500-6.

Dubois, Bruno, Howard H. Feldman, Claudia Jacova, Steven T. Dekosky, Pascale Barberger-Gateau, Jeffrey Cummings, André Delacourte, et al. 2007. "Research criteria for the diagnosis of Alzheimer's disease: revising the NINCDS–ADRDA criteria." *Lancet Neurol, 6(8)* 734-746. Accessed July 27, 2021. https://doi.org/10.1016/s1474-4422(07)70178-3.

Easton, Douglas F., Deborah Ford, D. Timothy Bishop, and Breast Cancer Linkage Consortium. 1995. "Breast and ovarian cancer incidence in BRCA1-mutation carriers. Breast Cancer Linkage Consortium." *Am. J. Hum. Genet. 56* 265-271.

Eckel, Robert H., Scott M. Grundy, and Paul Z. Zimmet. 2005. "The metabolic syndrome." *Lancet, 365(9468)* 1415-1428. Accessed July 26, 2021. https://doi.org/10.1016/S0140-6736(05)66378-7.

Ecker, Christina, and Margit Cichna-Markl. 2012. "Development and validation of a sandwich ELISA for the determination of potentially allergenic lupine in food." *Food Chemistry, 130(3)* 759-766. Accessed July 26, 2021. doi: 10.1016/j.foodchem.2011.07.100.

El-Agnaf, Omar M. A., Sultan A. Salem, Katerina E. Paleologou, Martin D. Curran, Mark J. Gibson, Jennifer A. Court, Michael G. Schlossmacher, and David Allsop. 2006. "Detection of oligomeric forms of alpha-synuclein protein in human plasma as a potential biomarker for Parkinson's disease." *Faseb J., 20(3)* 419-425. Accessed July 26, 2021. https://doi.org/10.1096/fj.03-1449com.

Elegado, Francisco B., Shara Mae T. Colegio, Vanessa Marie T. Lim, Andrea Therese R. Gervasio, Maria Teresa M. Perez, Marilen P. Balolong, Charina Grace B. Banaay, and Bernadette C. Mendoza. 2016. "Ethnic fermented foods of the Philippines with reference to lactic acid bacteria and yeasts." In *Ethnic Fermented Foods and Alcoholic Beverages of Asia*, by Jyoti Prakash Tamang, 323-340. New York: Springer Publishing.

Ellens, Daniel J., and Arno L. J. Gielkens. 1980. "A simple method for the purification of 5-aminosalicylic acid. Application of the product as substrate in enzyme-linked immunosorbent assay (ELISA)." *Journal of Immunological Methods, 37(3–4)* 325-332. Accessed July 26, 2021. https://doi.org/10.1016/0022-1759(80)90318-x.

Engvall, Eva. 2010. "The ELISA, enzyme-linked immunosorbent assay." *Clinical Chemistry, 56(2)* 319-320. Accessed July 26, 2021. https://doi.org/10.1373/clinchem.2009.127803.

Engvall, Eva, and Peter Perlmann. 1971. "Enzyme-linked immunosorbent assay (ELISA) quantitative assay of immunoglobulin G." *Immunochemistry, 8(9)* 871-874. Accessed July 27, 2021. https://doi.org/10.1016/0019-2791(71)90454-X.

Fæste, Christiane K., and Christin Plassen. 2008. "Quantitative sandwich ELISA for the determination of fish in foods." *Journal of*

Immunological Methods, 329(1-2) 45-55. Accessed July 25, 2021. https://doi.org/10.1016/j.jim.2007.09.007.

Figueroa, Lilian E., and Diego B. Genovese. 2019. "Fruit jellies enriched with dietary fibre: Development and characterization of a novel functional food product." *LWT, 111* 423-428. Accessed July 25, 2021. https://doi.org/10.1016/j.lwt.2019.05.031.

Fijan, Sabina. 2014. "Microorganisms with claimed probiotic properties: An overview of recent literature." *International Journal of Environmental Research and Public Health, 11(5)* 4745-4767. Accessed July 26, 2021. https://doi.org/10.3390/ijerph110504745.

Florowska, A., K. Krygier, T. Florowski, and E. Dłużewska. 2016. "Prebiotics as functional food ingredients preventing diet-related diseases." *Food and Function, 7(5)* 2147-2155. Accessed July 24, 2021. https://doi.org/10.1039/c5fo01459j.

Foltz, Martin, Patrick Ansems, Jessica Schwarz, Maria C. Tasker, Afrodite Lourbakos, and Cindy C. Gerhardt. 2008. "Protein hydrolysates induce CCK release from enteroendocrine cells and act as partial agonists of the CCK1 receptor." *Journal of Agricultural and Food Chemistry, 56(3)* 837-843. Accessed July 26, 2021. https://doi.org/10.1021/jf072611h.

Fredman, P., A. Wallin, K. Blennow, P. Davidsson, C. G. Gottfries, and L. Svennerholm. 1992. "Sulfatide as a biochemical marker in cerebrospinal fluid of patients with vascular dementia." *Acta Neurologica Scandinavica, 85(2)* 103-106. Accessed July 25, 2021. https://psycnet.apa.org/doi/10.1111/j.1600-0404.1992.tb04006.x.

Galarini, Roberta, Francesca Diana, Simone Moretti, Barbara Puppini, Giorgio Saluti, and Lidija Persic. 2014. "Development and validation of a new qualitative ELISA screening for multiresidue detection of sulfonamides in food and feed." *Food Control, 35(1)* 300-310. Accessed July 27, 2021. https://doi.org/10.1016/j.foodcont.2013.07.014.

Gan, Stephanie D., and Kruti R. Patel. 2013. "Enzyme immunoassay and enzyme-linked immunosorbent assay." *Journal of Investigative*

Dermatology, 133(9) 1-3. Accessed July 25, 2021. https://doi.org/10.1038/jid.2013.287.

Ganguli, Mary, Deborah Blacker, Dan G. Blazer, Igor Grant, Dilip V. Jeste, Jane S. Paulsen, Ronald C. Petersen, and Perminder S. Sachdev. 2011. "Classification of neurocognitive disorders in DSM-5: A work in progress." *American Journal of Geriatric Psychiatry, 19(3)* 205-210. Accessed July 26, 2021. https://doi.org/10.1097/JGP.0b013e3182051ab4.

Garrote, Graciela L., Maria A. Serradell, Analia G. Abraham, Maria C. Añon, Carlos A. Fossati, and Graciela L. De Antoni. 2005. "Development of an immunochemical method to detect Lactobacillus kefir." *Food and Agricultural Immunology, 16(3)* 221-233. Accessed July 23, 2021. https://doi.org/10.1080/09540100500244146.

German, Bruce, Eduardo J. Schiffrin, Roberto Reniero, Beat Mollet, Andrea Pfeifer, and Jean-Richard Neeser. 1999. "The development of functional foods: lessons from the gut." *Trends in Biotechnology, 17(12)* 492-499. Accessed July 24, 2021. https://doi.org/10.1016/S0167-7799(99)01380-3.

Goossens, Nicolas, Shigeki Nakagawa, Xiaochen Sun, and Yujin Hoshida. 2015. "Cancer biomarker discovery and validation." *Translational Cancer Research, 4(3)* 256-269. Accessed July 23, 2021. https://doi.org/10.3978/j.issn.2218-676X.2015.06.04.

Gorno-Tempini, M. L., A. E. Hillis, S. Weintraub, A. Kertesz, M. Mendez, S. F. Cappa, J. M. Ogar, et al. 2011. "Classification of primary progressive aphasia and its variants." *Neurology, 76(11)* 1006-1014. Accessed July 25, 2021. https://dx.doi.org/10.1212%2FWNL.0b013e31821103e6.

Gujral, Naiyana, Heejoo Yoo, Fatemeh Bamdad, Joo-Won Suh, and Hoon Sunwoo. 2017. "Sensitive double antibody sandwich ELISA for the quantification of phosvitin." *Food and Agricultural Immunology, 28(5)* 834-847. Accessed July 24, 2021. https://doi.org/10.1080/09540105.2017.1313821.

Guo, Jiebiao, Wangpei Liu, Xianquan Lan, Hualong Chen, and Zijun Xiao. 2016. "Development and evaluation of an immunochromatographic strip for rapid screening of sildenafil-type compounds as illegal additives in functional foods." *Food Additives & Contaminants: Part A, 33(7)* 1095-1104. Accessed July 26, 2021. https://doi.org/10.1080/19440049.2016.1203072.

Guzel-Seydim, Zeynep B., Tugba Kok-Tas, Annel K. Greene, and Atif C. Seydim. 2011. "Review: Functional properties of kefir." *Critical Reviews in Food Science and Nutrition, 51(3)* 261-268. Accessed July 25, 2021. https://doi.org/10.1080/10408390903579029.

Haghighi, Hamid R., Jianhua Gong, Carlton L. Gyles, M. Anthony Hayes, Babak Sanei, Payvand Parvizi, Haris Gisavi, James R. Chambers, and Shayan Sharif. 2005. "Modulation of Antibody-Mediated Immune Response by Probiotics in Chickens." *Clinical and Diagnostic Laboratory Immunology, 12(12)* 1387-1392. Accessed July 26, 2021. https://dx.doi.org/10.1128%2FCDLI.12.12.1387-1392.2005.

Hall, Jeff M., Ming K. Lee, Beth Newman, Jan E. Morrow, Lee A. Anderson, Bing Huey, and Mary-Claire King. 1990. "Linkage of early-onset familial breast cancer to chromosome 17q21." *Science, 250(4988)* 1684-1689.

Hampel, Harald, Katharina Buerger, Raymond Zinkowski, Stefan J. Teipel, Alexander Goernitz, Niels Andreasen, Magnus Sjoegren, et al. 2004. "Measurement of phosphorylated tau epitopes in the differential diagnosis of Alzheimer disease: a comparative cerebrospinal fluid study." *Arch Gen Psychiatry, 61(1)* 95-102. Accessed July 27, 2021. https://doi.org/10.1001/archpsyc.61.1.95.

Hasler, Clare M. 2002. "Functional foods: Benefits, concerns and challenges - A position paper from the American Council on Science and Health." *The Journal of Nutrition, 132(12)* 3772-3781. Accessed July 24, 2021. https://doi.org/10.1093/jn/132.12.3772.

Henry, C. J. 2010. "Functional foods." *European Journal of Clinical Nutrition, 64* 657-659. Accessed July 24, 2021. https://doi.org/10.1038/ejcn.2010.101.

Henry, N. Lynn, and Daniel F. Hayes. 2012. "Cancer biomarkers." *Molecular Oncology, 6(2)* 140-146. Accessed July 25, 2021. https://doi.org/10.1016/j.molonc.2012.01.010.

Hill, James O., and John C. Peters. 2002. "Biomarkers and functional foods for obesity and diabetes." *British Journal of Nutrition, 88(S2)* S213-S218. Accessed July 26, 2021. https://doi.org/10.1079/bjn2002685.

Holzhauser, Thomas, and Stefan Vieths. 1999. "Indirect competitive ELISA for determination of traces of peanut (*Arachis hypogaea* L.) protein in complex food matrices." *Journal of Agricultural and Food Chemistry, 47(2)* 603-611. Accessed July 23, 2021. https://doi.org/10.1021/jf980775f.

Hornbeck, Peter. 1992. "Enzyme-Linked Immunosorbent Assays." *Current Protocols in Immunology, 1(1)* 2.1.1-2.1.22. Accessed July 23, 2021. https://doi.org/10.1002/0471142735.im0201s01.

Hosseini, Samira, Patricia Vázquez-Villegas, M. Rito-Palomares, and Sergio O. Martinez. 2018. "Enzyme-Linked Immunosorbent Assay (ELISA)." In *Springer Briefs in Applied Sciences and Technology*, by Samira Hosseini, Patricia Vázquez-Villegas, M. Rito-Palomares and Sergio O. Martinez, 67-115. Berlin: Springer-Verlag.

Imperiale, Thomas F., David F. Ransohoff, Steven H. Itzkowitz, Theodore R. Levin, Philip Lavin, Graham P. Lidgard, David A. Ahlquist, and Barry M. Berger. 2014. "Multitarget stool DNA testing for colorectal-cancer screening." *The New England Journal of Medicine, 371(2)* 187-188. Accessed July 26, 2021. doi: 10.1056/NEJMoa1311194.

Jensen, J. L., G. D. Maclean, M. R. Suresh, A. Almeida, D. Jette, S. Lloyd, D. Bodnar, M. Krantz, and B. M. Longenecker. 1991. "Possible utility of serum determinations of CA 125 and CA 27.29 in breast cancer management." *The International Journal of Biological Markers, 6(1)* 1-6.

Johnson, P. J., and Roger Williams. 1980. "Serum alpha-fetoprotein estimations and doubling time in hepatocellular carcinoma: influence of therapy and possible value in early detection." *Journal of the*

National Cancer Institute, 64(6) 1329-1332. Accessed July 27, 2021. https://doi.org/10.1093/jnci/64.6.1329.

Joshi, Aditya, Venkatesh Basappa Krishnamurthy, Heather Purichia, Laura Hollar-Wilt, Edward Bixler, and Mark Rapp. 2012. "“What’s in a name?” Delirium by any other name would be as deadly. A review of the nature of delirium consultations." *J. Psychiatr. Pract., 18(6)* 413-418. Accessed July 26, 2021. https://doi.org/10.1097/01.pra.0000422739.49377.17.

Kamel, Hala Fawzy Mohamed, and Hiba Saeed Bagader Al-Amodi. 2016. "Cancer biomarkers." In *Role of Biomarkers in Medicine*, by Mu Wang and Frank Witzmann, 1-32. IntechOpen.

Kantarijan, Hagop, Charles Sawyers, Andreas Hochhaus, Francois Guilhot, Charles Schiffer, Carlo Gambacorti-Passerini, Dietger Niederwieser, et al. 2002. "Hematologic and cytogenetic responses to imatinib mesylate in chronic myelogenous leukemia." *Engl J Med., 346(9)* 645-652. Accessed July 27, 2021. https://doi.org/10.1056/nejmoa011573.

Karelakis, Christos, Panagiotis Zevgitis, Konstantinos Galanopoulos, and Konstadinos Mattas. 2020. "Consumer Trends and Attitudes to Functional Foods." *Journal of International Food and Agribusiness Marketing, 32(3)* 266-294. Accessed July 27, 2021. https://doi.org/10.1080/08974438.2019.1599760.

Kato, Kanefusa, Yoshitaka Hamaguchi, Sachiyo Okawa, Eiji Ishikawa, Keiko Kobayashi, and Nobuhiko Katunuma. 1977. "Use of rabbit antibody IgG bound onto plain and aminoalkylsilyl glass surface for the enzyme-linked sandwich immunoassay." *The Journal of Biochemistry, 82(1)* 261-266. Accessed July 25, 2021. https://doi.org/10.1093/oxfordjournals.jbchem.a131678.

Kazemi, Maryam Ghayoor, and Javad Feizy. 2020. "Overview of the importance of ELISA technique and application in food industry." *NAISL, 4(4)* 19-25.

Kibar, Yusuf, Serdar Goktas, Selim Kilic, Halil Yaman, Onder Onguru, and Ahmet Fuat Peker. 2006. "Prognostic value of cytology, nuclear matrix protein 22 (NMP22) test, and urinary bladder cancer II (UBC

II) test in early recurrent transitional cell carcinoma of the bladder." *Annals of Clinical and Laboratory Science, 36(1)* 31-38.

Kitashoji, Emi, Nobuo Koizumi, Talitha Lea V. Lacuesta, Daisuke Usuda, Maricel R. Ribo, Edith S. Tria, Winston S. Go, et al. 2015. "Diagnostic accuracy of recombinant immunoglobulin-like protein A-based IgM ELISA for the early diagnosis of leptospirosis in the Philippines." *PLoS Neglected Tropical Diseases, 9(6):e0003879* Accessed July 26, 2021. https://doi.org/10.1371/journal.pntd.0003879.

Kiyota, Kyohei, Masato Yoshimitsu, Taro Satsuki-Murakami, Kazuhiko Akutsu, Keiji Kajimura, and Tetsuo Yamano. 2017. "Detection of the tomato allergen Sola l 1 and evaluation of its reactivity after heat and papain treatment." *Food and Agricultural Immunology, 28(6)* 1450-1459. Accessed July 24, 2021. https://doi.org/10.1080/09540105.2017.1347914.

Koenen, Marjorie E., Remy van der Hulst, Mariska Leering, Suzan H. M. Jeurissen, and Wim J. A. Boersma. 2004. "Development and validation of a new in vitro assay for selection of probiotic bacteria that express immune-stimulating properties in chickens in vivo." *FEMS Immunology & Medical Microbiology, 40(2)* 119-127. Accessed July 26, 2021. https://doi.org/10.1016/s0928-8244(03)00306-7.

Koizumi, F., H. Odagiri, H. Fujimoto, T. Kawamura, and A. Ishimori. 1992. "Clinical evaluation of four tumor markers (CEA, TPA, CA50 and CA72-4) in colorectal cancer." *Rinsho Byori: The Japanese Journal of Clinical Pathology, 40(5)* 523-528.

Konstantinidi, Melina, and Antonios E. Koutelidakis. 2019. "Functional foods and bioactive compounds: A review of its possible role on weight management and obesity's metabolic consequences." *Medicines, 6(3):94* Accessed July 26, 2021. https://doi.org/10.3390/medicines6030094.

Konstantinou, George N. 2017. "Enzyme-linked immunosorbent assay (ELISA)." *Methods in Molecular Biology, 1592* 79-94. Accessed July 25, 2021. https://doi.org/10.1007/978-1-4939-6925-8_7.

Kotecha, Dipak, Marcus D. Flather, Dan Atar, Peter Collins, John Pepper, Elizabeth Jenkins, Christopher M. Reid, and David Eccleston. 2019. "B-type natriuretic peptide trumps other prognostic markers in patients assessed for coronary disease." *BMC Medicine, 17(1)* 1-11. Accessed July 23, 2021. https://doi.org/10.1186/s12916-019-1306-9.

Koutelidakis, Antonios E., and Charalampia Dimou. 2016. "The effects of functional food and bioactive compounds on biomarkers of cardiovascular diseases." In *Functional foods for chronic diseases (1st ed.)*, by D. Martirosyan, 89-117. Dallas: Food Science Publisher.

Kurtzman, James T., H. Wilson, and Ch. V. Rao. 2001. "A proposed role for hCG in clinical obstetrics." *Seminars in Reproductive Medicine, 19(1)* 063-068. Accessed July 24, 2021. doi: 10.1055/s-2001-13912.

Lacorn, Markus, Tina Dubois, Susanne Siebeneicher, and Thomas Weiss. 2016. "Accurate and sensitive quantification of soy proteins in raw and processed food by sandwich ELISA." *Food Science and Technology, 4(4)* 69-77. Accessed July 27, 2021. http://dx.doi.org/10.13189/fst.2016.040404.

Lattanzio, Francesca, Samir Abu-Rumeileh, Alessia Franceschini, Hideaki Kai, Giulia Amore, Ilaria Poggiolini, Marcello Rossi, et al. 2017. "Prion-specific and surrogate CSF biomarkers in Creutzfeldt-Jakob disease: diagnostic accuracy in relation to molecular subtypes and analysis of neuropathological correlates of p-tau and Aβ42 levels." *Acta Neuropathol., 133(4)* 559-578. Accessed July 26, 2021. https://doi.org/10.1007/s00401-017-1683-0.

Lee, Ji-Yun, and Chang Jong Kim. 2010. "Determination of allergenic egg proteins in food by protein-, mass spectrometry-, and DNA-based methods." *Journal of AOAC International, 93(2)* 462-477.

Legg, Timothy J. 2018. *Neurocognitive Disorders (Organic Brain Syndrome).* September 4. Accessed July 25, 2021. https://www.healthline.com/health/organic-brain-syndrome.

Leitão, M. J., I. Baldeiras, M. R. Almeida, M. H. Ribeiro, A. C. Santos, M. Ribeiro, J. Tomás, S. Rocha, I. Santana, and C. R. Oliveira. 2016. "Sporadic Creutzfeldt-Jakob disease diagnostic accuracy is improved by a new CSF ELISA 14-3-3?? assay." *Neuroscience, 322* 398-407. Accessed July 24, 2021. https://doi.org/10.1016/j.neuroscience.2016.02.057.

Li, Katherine J., Elske M. Brouwer-Brolsma, Kathryn J. Burton-Pimentel, Guy Vergères, and Edith J. M. Feskens. 2021. "A systematic review to identify biomarkers of intake for fermented food products." *Genes & Nutrition, 16(5)* 1-17. Accessed July 24, 2021. https://doi.org/10.1186/s12263-021-00686-4.

Li, Meng, Haiping Su, Yan Tu, Yazhuo Shang, Yu Liu, Changjun Peng, and Honglai Liu. 2019. "Development and application of an efficient medium for chromogenic catalysis of tetramethylbenzidine with horseradish peroxidase." *ACS Omega, 4(3)* 5459-5470. Accessed July 24, 2021. https://doi.org/10.1021/acsomega.8b03345.

Lilamand, M., C. Hourregue, and C. Paquet. 2020. "Interest of biological biomarkers in the diagnostic approach of neurocognitive disorders in the elderly." *Revue Neurologique, 176(9)* 677-683. Accessed July 23, 2021. https://doi.org/10.1016/j.neurol.2019.12.006.

Lin, Alice V. 2015. "Chapter 5: Indirect ELISA." In *ELISA: Methods and Protocols, Methods in Molecular Biology (Vol. 1318)*, by Robert Hnasko, 51-59. New York: Springer Science + Business Media.

Linares, Daniel M., Carolina Gómez, Erica Renes, José M. Fresno, María E. Tornadijo, R. P. Ross, and Catherine Stanton. 2017. "Lactic Acid Bacteria and Bifidobacteria with Potential to Design Natural Biofunctional Health-Promoting Dairy Foods." *Frontiers in Microbiology* Accessed July 24, 2021. https://doi.org/10.3389/fmicb.2017.00846 .

Llorens, Franc, Niels Kruse, Matthias Schmitz, Nadine Gotzmann, Ewa Golanska, Katrin Thüne, Orgeta Zejneli, et al. 2017. "Evaluation of α-synuclein as a novel cerebrospinal fluid biomarker in different forms of prion diseases." *Alzheimer's Dement., 13(6)* 710-719. Accessed July 26, 2021. https://doi.org/10.1016/j.jalz.2016.09.013.

London, Calvin. 2010. "Functional foods that boost the immune system." In *Functional food product development*, by Jim Smith and Edward Charter, 295-321. Blackwell Publishing Ltd.

Lopez, Anna Lena, Cameron Adams, Michelle Ylade, Ramesh Jadi, Jedas Veronica Daag, Caitlyn T. Molloy, Kristal An Agrupis, et al. 2021. "Determining dengue virus serostatus by indirect IgG ELISA compared with focus reduction neutralisation test in children in Cebu, Philippines: A prospective population-based study." *The Lancet Global Health, 9(1)* e44-e51. Accessed July 24, 2021. https://doi.org/10.1016/S2214-109X(20)30392-2.

Lu, Xiaonan. 2017. *Sensing Techniques for Food Safety and Quality Control.* London: Royal Society of Chemistry.

Manila Bulletin. 2021. *DOST-FNRI to produce functional food products with prebiotic properties.* News, Manila: Manila Bulletin.

Luckheeram, Rishi Vishal, Rui Zhou, Asha Devi Verma, and Bing Xia. 2012. "CD4 + T Cells: Differentiation and Functions." *Clinical and Developmental Immunology, 2012:925135* Accessed July 26, 2021. https://doi.org/10.1155/2012/925135.

Ludwig, Joseph A., and John N. Weinstein. 2005. "Biomarkers in cancer staging, prognosis and treatment selection." *Nature Reviews Cancer, 5* 845-856. Accessed July 25, 2021. https://doi.org/10.1038/nrc1739.

Luvián-Morales, Julissa, Francisco Omar Varela-Castillo, Laura Flores-Cisneros, Lucely Cetina-Pérez, and Denisse Castro-Eguiluz. 2021. "Functional foods modulating inflammation and metabolism in chronic diseases: a systematic review." *Ciritical Reviews in Food Science and Nutrition* 1-22. Accessed July 23, 2021. https://doi.org/10.1080/10408398.2021.1875189.

Maartens, Gary, Connie Celum, and Sharon R. Lewin. 2014. "HIV infection: Epidemiology, pathogenesis, treatment, and prevention." *The Lancet, 384(9939)* 258-271. Accessed July 24, 2021. https://doi.org/10.1016/s0140-6736(14)60164-1.

Maeda, Kenji, Debananda Das, Takuya Kobayakawa, Hirokazu Tamamura, and Hiroaki Takeuchi. 2019. "Discovery and Development of Anti-HIV Therapeutic Agents: Progress Towards Improved HIV Medication." *Current Topics in Medicinal Chemistry, 19(18)* 1621-1649. Accessed July 24, 2021. https://doi.org/10.2174/1568026619666190712204603.

Martin, Archer K., Anita K. Malhotra, Breandan L. Sullivan, and Harish Ramakrishna. 2016. "Troponin elevations in patients with chronic cardiovascular disease: An analysis of current evidence and significance." *Annals of Cardiac Anaesthesia, 19(2)* 321-327. Accessed July 25, 2021. https://doi.org/10.4103/0971-9784.179638.

Matsuda, H., H. Tanaka, B. L. Blas, J. S. Noseñas, T. Tokawa, and S. Ohsawa. 1984. "Evaluation of ELISA with ABTS, 2-2'-azino-di-(3-ethylbenzthiazoline sulfonic acid), as the substrate of peroxidase and its application to the diagnosis of schistosomiasis." *The Japanese Journal of Experimental Medicine, 54(3)* 131-138.

Matsuoka, Y. 1990. "Basic and clinical aspects of tumor markers–with special reference to CEA." *Rinsho Byori: The Japanese Journal of Clinical Pathology, 38(1)* 31-34.

Mayeux, Richard. 2004. "Biomarkers: Potential uses and limitations." *Neurotherapeutics: The Journal of the American Society for Experimental Neurotherapeutics, 1(2)* 182-188. Accessed July 25, 2021. https://dx.doi.org/10.1602%2Fneurorx.1.2.182.

Mazzaferri, E. L., R. J. Robbins, C. A. Spencer, L. E. Braverman, F. Pacini, L. Wartofsky, B. R. Haugen, et al. 2003. "A consensus report of the role of serum thyroglobulin as a monitoring method for low-risk patients with papillary thyroid carcinoma." *The Journal of Clinical Endocrinology and Metabolism, 88(4)* 1433-1441. Accessed July 24, 2021. https://doi.org/10.1210/jc.2002-021702.

McDonough, Ian M., and Rebecca S. Allen. 2019. "Biological markers of aging and mental health: A seed and soil model of neurocognitive disorders." *Aging and Mental Health, 23(7)* 793-799. Accessed July 23, 2021. https://doi.org/10.1080/13607863.2018.1531383.

McGuire, W. L., K. B. Horwitz, O. H. Pearson, and A. Segaloff. 1977. "Current status of estrogen and progesterone receptors in breast cancer." *Cancer, 39(6 Suppl)* 2934-2947. Accessed July 24, 2021. https://doi.org/10.1002/1097-0142(197706)39:6%3C2934::aid-cncr2820390680%3E3.0.co;2-p.

McKeith, I. G., D. W. Dickson, J. Lowe, M. Emre, J. T. O'Brien, H. Feldman, J. Cummings, et al. 2005. "Diagnosis and management of dementia with Lewy bodies: third report of the DLB Consortium." *Neurology, 65(12)* 1863-1872. Accessed July 25, 2021. https://doi.org/10.1212/01.wnl.0000187889.17253.b1.

Minami, Tatsuya, Ryosuke Tateishi, Masyuko Kondo, Ryo Nakagomi, Naoto Fujiwara, Masaya Sato, Koji Uchino, et al. 2015. "Serum alpha-fetoprotein has high specificity for the early detection of hepatocellular carcinoma after hepatitis C virus eradication in patients." *Medicine, 94(23):e901* Accessed July 25, 2021. https://dx.doi.org/10.1097%2FMD.0000000000000901.

Mirmiran, Parvin, Zahra Bahadoran, and Fereidoun Azizi. 2014. "Functional foods-based diet as a novel dietary approach for management of type 2 diabetes and its complications: A review." *World Journal of Diabetes, 5(3)* 267-281. Accessed July 25, 2021. https://dx.doi.org/10.4239%2Fwjd.v5.i3.267.

Molina, Rafael, Vivian Barak, Arie van Dalen, Michael J. Duffy, Roland Einarsson, Massimo Gion, Helenka Goike, et al. 2005. "Tumor markers in breast cancer-European group on tumor markers recommendations." *Tumour Biology: The Journal of the International Society for Oncodevelopmental Biology and Medicine, 26(6)* 281-293. Accessed July 23, 2021. https://doi.org/10.1159/000089260.

Montine, T. J., M. F. Beal, D. Robertson, M. E. Cudkowicz, I. Biaggioni, H. O'Donnell, W. E. Zackert, L. J. Roberts, and J. D. Morrow. 1999. "Cerebrospinal fluid F2-isoprostanes are elevated in Huntington's disease." *Neurology, 52(5)* 1104-1105. Accessed July 25, 2021. https://doi.org/10.1212/WNL.52.5.1104.

Moriyama, Tatsuya, Erika Yano, Nobuhiro Zaima, Keiko Miyazaki, Kou Shirotsuki, Aki Sato, and Makoto Sawaguchi. 2020. "Development of pretreatment protocols for determination of soybean β-conglycinin in processed soybean foods using commercial ELISA kits." *Journal of Nutritional Science and Vitaminology, 66(3)* 270-277. Accessed July 26, 2021. https://doi.org/10.3177/jnsv.66.270.

Na, Muzi, Nan Dou, Naiwen Ji, Dixin Xie, Jie Huang, Katherine L. Tucker, and Xiang Gao. 2020. "Food Insecurity and Cognitive

Function in Middle to Older Adulthood: A Systematic Review." *Advances in Nutrition, 11(3)* 667-676. Accessed July 24, 2021. https://doi.org/10.1093/advances/nmz122.

Nataraj, Basavaprabhu H., Syed Azmal Ali, Pradip V. Behare, and Hariom Yadav. 2020. "Postbiotics-parabiotics: The new horizons in microbial biotherapy and functional foods." *Microbial Cell Factories, 19:168* Accessed July 23, 2021. https://doi.org/10.1186/s12934-020-01426-w.

National Cancer Institute. 2021. *What is Cancer?* https://www.cancer.gov/about-cancer/understanding/what-is-cancer.

National Center for Biotechnology Information. n.d. *PubChem Compound Summary for CID 5333, Sulfanilamide.* Accessed June 20, 2021. https://pubchem.ncbi.nlm.nih.gov/compound/Sulfanilamide.

—. n.d. *PubChem Compound Summary for CID 5430, Thiabendazole.* Accessed June 20, 2021. https://pubchem.ncbi.nlm.nih.gov/compound/Thiabendazole.

Hill, Colin, Francisco Guarner, Gregor Reid, Glenn R. Gibson, Daniel J. Merenstein, Bruno Pot, Lorenzo Morelli, et al. 2014. "The International Scientific Association for Probiotics and Prebiotics consensus statement on the scope and appropriate use of the term probiotic." *Nature Reviews Gastroenterology & Hepatology, 11* 506-514. Accessed July 23, 2021. https://doi.org/10.1038/nrgastro.2014.66.

Nishino, H., H. Tokuda, Y. Satomi, Mitsuharu Masuda, Ping Bu, M. Onozuka, S. Yamaguchi, et al. 1999. "Cancer prevention by carotenoids." *Pure and Applied Chemistry, 71* 2273-2278.

O'Brien, Jacob, Heyam Hayder, Yara Zayed, and Chun Peng. 2018. "Overview of microRNA biogenesis, mechanisms of actions, and circulation." *Frontiers in Endocrinology* 1-12. Accessed July 26, 2021. https://doi.org/10.3389/fendo.2018.00402.

Oh, M., J. H. Lee, S. A. Kim, and M. Kim. 2018. "Analytical method development for determination of silymarin by LC-MS/MS for related health functional foods." *Journal of Food Hygiene and Safety,*

33(2) 124-130. Accessed July 26, 2021. doi: 10.13103/jfhs.2018.33.2.124.

Omar, Sharaf S., Moawiya A. Haddad, and Salvatore Parisi. 2020. "Validation of HPLC and enzyme-linked immunosorbent assay (ELISA) techniques for detection and quantification of aflatoxins in different food samples." *Foods, 9(5):661* Accessed July 23, 2021. https://doi.org/10.3390/foods9050661.

Ovchinnikova, Ekaterina S., Daniela Schmitter, Eline L. Vegter, Jozine M. Ter Maaten, Mattia A. E. Valente, Licette C. Y. Liu, Pim van der Harst, et al. 2016. "Signature of circulating microRNAs in patients with acute heart failure." *European Journal of Heart Failure, 18(4)* 414-423. Accessed July 27, 2021. https://doi.org/10.1002/ejhf.332.

Paik Soonmyung, Gong Tang, Steven Shak, Chungyeul Kim, Joffre Baker, Wanseop Kim, Maureen Cronin, Frederick L. Baehner, Drew Watson, John Bryant, Joseph P. Constatino, Charles E. Geyer Jr., D Lawrence Wickerham, and Norman Wolmark. (2006). "Gene Expression and benefit of chemotherapy in women with Node-Negative, Estrogen Receptor-Positive Breast Cancer. *Journal of Clinical Oncology*, 24: 3726–34.

Paik Soonmyung, Gong Tang, Steven Shak, Chungyeul Kim, Joffre Baker, Maureen Cronin, Frederick L. Baehner, Michael G. Walker, Drew Watson, Taesung Park, William Hiller, Edwin R. Fisher, D Lawrence Wickerham, John Bryant, and Normal Wolmark. (2004). A Multigene Assay to Predict Recurrence of Tamoxifen-Treated, Node-Negative Breast Cancer. *The New England Journal of Medicine*, 351: 2817–2826.

Palou, A., Serra, F., and Pico, C. (2003). "General Aspects on the Assessment of Functional Foods in the European Union." *European Journal of Clinical Nutrition,* 57(S1): S12–S17. Accessed July 3, 2021. doi: 10.1038/sj.ejcn.1601822.

Pao William, and Nicolas Girard. (2011). "New Driver Mutations in Non-Small-Cell Lung Cancer. *The Lancet Oncology,* 12(2): 175–180.

Park, Kyung Chan, David C. Gaze, Paul O. Collinson, and Michael S. Marber. (2017). "Cardiac Troponins: From Myocardial Infarction to

Chronic Disease." *Cardiovascular Research,* 113(14): 1708–1718. Accessed July 10, 2021. doi.org/10.1093/cvr/cvx183.

Pastor, Lucía, Aina Casellas, Jorge Carrillo, Sergi Alonso, Erica Parker, Laura Fuente-Soro, Chenjerai Jairoce, Inacio Mandomando, Julià Blanco, and Denise Naniche. (2017). "IP-10 Levels as an Accurate Screening Tool to Detect Acute HIV Infection in Resource-Limited Settings." *Scientific Reports, 7*(1): 1–9. Accessed July 14, 2021. doi.org/10.1038/s41598-017-08218-0.

Pathak, Manju, and Danik M. Martirosyan. (2011). "Immunodetection and quantification of Insulin-Like Antigens in Sprouts: Development of an Efficient Functional Food." *Functional Foods in Health and Disease,* 1(11): 492–507.

Patrick, Weber, Steinhart Hans, and Paschke Angelika. (2009). Determination of the Bovine Food Allergen Casein in White Wines by Quantitative Indirect ELISA, SDS-PAGE, Western Blot and Immunostaining. *Journal of Agricultural and Food Chemistry,* 57: 8399–8405.

Pelaez-Lorenzo, Cristina, Jose Carlos Diez-Mesa, Isabel Vasallo, and Mercedes de Frutos. (2010). "Development of an Optimized ELISA and a Sample Preparation Method for the Detection of β-Lactoglobulin Traces in Baby Foods." *Journal of Agricultural and Food Chemistry*, 58(3): 1664–1671. Accessed July 9, 2021. doi.org/ 10.1021/jf9041485.

Peng, Mengfei, Zajeba Tabashsum, Mary Anderson, Andy Truong, Ashley K. Houser, Joselyn Padilla, Ahlam Akmel, Jacob Bhatti, Shaik O. Rahaman, and Debabrata Biswas. (2020). "Effectiveness of Probiotics, Prebiotics, and Prebiotic-Like Components in Common Functional Foods." *Comprehensive Reviews in Food Science and Food Safety*, 19(4): 1908–1933. Accessed July 3, 2021. doi.org/ 10.1111/1541-4337.12565.

Petersen, Ronald C., Rosebud O. Robers, David S. Knopman, Bradley F. Boeve, Yonas E. Geda, Robert J. Ivnik, Glenne E. Smith, and Clifford R. Jack Jr. (2009). "Mild Cognitive Impairment: Ten Years Later." *Archives of Neurology*, 66: 1546–54.

Piccart-Gebhart, Martine J. et al., (2005). "Trastuzumab After Adjuvant Chemotherapy in HER2-positive Breast Cancer." *The New England Journal of Medicine,* 353(16): 1659–72.

Picó, Catalina, Francisca Serra, Ana María Rodríguez, Jaap Keijer, and Andreu Palou. (2019). "Biomarkers of Nutrition and Health: New Tools for New Approaches." *Nutrients,* 11(5): 1092. Accessed July 27, 2021. doi: 10.3390/nu11051092.

Planès, Rémi, Manutea Serrero, Kaoutar Leghmari, Lbachir BenMohamed, and Elmostafa Bahraoui. (2018). "HIV-1 Envelope Glycoproteins Induce the Production of TNF-α and IL-10 in Human Monocytes by Activating Calcium Pathway." *Scientific Reports*, 8(1): 1–15. Accessed July 25, 2021. doi.org/10.1038/s41598-018-35478-1.

Poulose, Shibu M., Marshall G. Miller, Tammy Scott, and Barbara Shukitt-Hale. (2017). "Nutritional Factors Affecting Adult Neurogenesis and Cognitive Function." *Advances in Nutrition*, 8(6): 804–811. Accessed July 15, 2021. doi.org/10.3945/an.117.016261.

Rad, Aziz Homayouni, Leili Aghebati Maleki, Hossein Samadi Kafil, Hamideh Fathi Zavoshti, and Aamin Abbasi. (2020). "Postbiotics as Novel Health-Promoting Ingredients in Functional Foods." *Health Promotion Perspectives*, 10(1): 3–4. Accessed July 15, 2021. doi.org/10.15171/hpp.2020.02.

Rafter, Joseph J. (2002). "Scientific Basis of Biomarkers and Benefits of Functional Foods for Reduction of Disease Risk: Cancer." *British Journal of Nutrition*, 88(S2): S219–S224. Accessed July 25, 2021. doi: 10.1079/bjn2002686.

Rascovsky, Katya, et al., (2011). "Sensitivity of Revised Diagnostic Criteria for the Behavioural Variant of Frontotemporal Dementia." *Brain*, 134: 2456–2477.

Rochlani, Yogita, Naga Venkata Pothineni, Swathi Kovelamudi, and Jawahar L. Mehta. (2017). "Metabolic Syndrome: Pathophysiology, Management, and Modulation by Natural Compounds." *Therapeutic Advances in Cardiovascular Disease*, 11(8): 215–225.

Roderburg, Christoph, and Tom Luedde. (2014). "Circulating

microRNAs as Markers of Liver Inflammation, Fibrosis and Cancer." *Journal of Hepatology*, 61(6): 1434–1437. Accessed July 15, 2021. doi.org/10.1016/j.jhep.2014.07.017.

Romond, Edward H., et al. (2005). "Trastuzumab Plus Adjuvant Chemotherapy for Operable HER2-Positive Breast Cancer." *The New England Journal of Medicine,* 353(16): 1673–84.

Rosado-Sánchez, Isaac, et al. (2017). "Higher Levels of IL-6, CD4 Turnover and Treg Frequency are Already Present Before cART in HIV-Infected Subjects with Later Low CD4 Recovery." *Antiviral Research*, 142: 76–82. Accessed July 10, 2021. doi.org/10.1016/j.antiviral.2017.03.015.

Ross, Jeffrey S., Elzbieta A. Slodkowska, W. Fraser Symmans, Lajos Pusztai, Peter M. Ravdin, and Gabriel N. Hortobagyi. (2009). "The HER-2 Receptor and Breast Cancer: Ten Years of Targeted Anti-HER-2 Therapy and Personalized Medicine." *Oncologist*, 14, 320–368.

Rutstein, Sarah E., et al. (2017). "Clinical and Public Health Implications of Acute and Early HIV Detection and Treatment: A Scoping Review. *Journal of the International AIDS Society*, 20(1): 1–13. Accessed July 20, 2021. doi.org/10.7448/IAS.20.1.21579.

Sachdev, Perminder S., Deborah Blacker, Dan G. Blazer, Mary Ganguli, Dilip V. Jeste, Jane S. Paulsen, and Ronald C. Petersen. (2014). "Classifying Neurocognitive Disorders: The DSM-5 Approach." *Nature Reviews Neurology,* 10(11): 634–642. Accessed July 9, 2021. doi.org/10.1038/nrneurol.2014.181.

Sacks, Frank M., et al. (2001). "Effects on Blood Pressure of Reduced Dietary Sodium and the Dietary Approaches to Stop Hypertension (DASH) Diet." *The New England Journal of Medicine*, 344(1): 3–10. Accessed July 10, 2021. doi.org/10.1056/nejm200101043440101.

Sahasrabudhe, Neha M., et al. (2018). "Dietary Fiber Pectin Directly Blocks Toll-Like Receptor 2–1 and Prevents Doxorubicin-Induced Ileitis." *Frontiers in Immunology,* 9(383), 1–19. Accessed July 10, 2021. doi: 10.3389/fimmu.2018.00383.

Sakamoto, Seiichi, Waraporn, Putalun, Sornkanok Vimolmangkang,

Waranyoo Phoolcharoen, Yukihiro Shoyama, Hiroyuki Tanaka, and Satoshi Morimoto. (2017). "Enzyme-Linked Immunosorbent Assay for the Quantitative/Qualitative Analysis of Plant Secondary Metabolites." *Journal of Natural Medicines*, 72(1), 32–42. Accessed July 3, 2021. doi.org/10.1007/s11418-017-1144-z.

Samanta, Saptadip. (2020). "Potential Impacts of Prebiotics and Probiotics in Cancer Prevention." *Anti-Cancer Agents in Medicinal Chemistry*, 21. Accessed July 20, 2021. doi.org/10.2174/1871520621999201210220442.

Sarojnalini, Chungkham, and Abdul Hei. (2019). "Fish as an Important Functional Food for Quality Life." In *Functional Foods*. IntechOpen. Accessed July 1, 2021. doi.org/10.5772/intechopen.81947.

Schneid, Andréa dos Santos, Kelly Lameiro Rodrigues, Davi Chemello, Eduardo Cesar Tondo, Marco Antônio Zacchia Ayub, and José Antonio Guimarães Alexio. (2006). "Evaluation of an Indirect ELISA for the Detection of *Salmonella* in Chicken Meat." *Brazilian Journal of Microbiology*, 37: 350–355.

Schneider, Keith R., Renée Goodrich Schneider, Rachael Silverberg, Ploy Kurdmongkoltham, and Bruna Bertoldi. (2021, June 20). *Preventing Foodborne Illness: Bacillus cereus.* Retrieved from National Institute of Food and Agriculture: https://nifa.usda.gov/sites/default/files/resource/Preventing-Foodborne-Illness-Bacillus-cereus.pdf.

Shah, Karishma, and Panagiotis Maghsoudlou. (2016). "Enzyme-Linked Immunosorbent Assay (ELISA): The Basics." *British Journal of Hospital Medicine*, 77(7): C98–C101.

Shaw, Alice T. et al. (2013). "Crizotinib Versus Chemotherapy in Advanced ALK-Positive Lung Cancer." *The New England Journal for Medicine*, 368(25): 2385–94.

Sheehan, Viven M., P. Ross, and Gerald F. Fitzgerald. (2007). "Assessing the Acid Tolerance and the Technological Robustness of Probiotic Cultures for Fortification in Fruit Juices. *Innovative Food Science and Emerging Technologies*, 8(2): 279–284. Accessed July 20, 2021. doi.org/10.1016/j.ifset.2007.01.007.

Sidhu, Jiwan S., Yearul Kabir, and Fatma G. Huffman. (2007). "Functional Foods from Cereal Grains." *International Journal of Food Properties*, 10(2): 231–244. Accessed July 20, 2021. doi.org/10.1080/10942910601045289.

Simmons, Rachel P., Eileen P. Scully, Erin E. Groden, Kelly F. Benedict, J. Judy Chang, Kim Lane, Jeff Lifson, Eric Rosenburg, Douglas A. Lauffenburger, and Marcus Altfeld. (2013). "HIV-1 Infection Induces Strong Production of IP-10 through TLR7/9-Dependent Pathways. *Aids*, 27(16): 2505–2517. Accessed July 25, 2021. doi.org/10.1097/01.aids.0000432455.06476.bc.

Simpson, Joseph R. (2014). "DSM-5 and neurocognitive disorders." *Journal of the American Academy of Psychiatry and the Law, 42*(2), 159–164.

Singh, Gurmit, Ligia Velasquez, Anne-Catherine Huet, Philippe Delahaut, Nathalie Gillard, and Terry Koerner. (2019). "Development of a Sensitive Polyclonal Antibody-Based Competitive Indirect ELISA for Determination of Citrinin in Grain-Based Foods." *Food Additives and Contaminants: Part A*. 36(10): 1567–1573.

Sjögren, Magnus, M. Blomberg, Michael Jonsson, Lars-Olof Wahlund, A. Edman, K. Lind, L. Rosengren, K. Blennow, and Asa K. Wallin. (2001). "Neurofilament Protein in Cerebrospinal Fluid: A Marker of White Matter Changes. *Journal of Neuroscience Research*, 66: 510–516.

Slattery, Conor, Paul D. Cotter, and Paul W. O'Toole. (2019). "Analysis of Health Benefits Conferred by *Lactobacillus* Species from Kefir. *Nutrients*, 11(6): 1252.

So-Armah, Kaku, and Matthew S. Freiberg. (2018). "HIV and Cardiovascular Disease: Update on Clinical Events, Special Populations, and Novel Biomarkers." *Current HIV/AIDS Reports*, 15(3), 233–244. Accessed July 20, 2021. doi.org/10.1007/s11904-018-0400-5.

Song, Y., Wang, Y. Y., Zhang, Y., and Wang, S. (2012). "Development of Enzyme-Linked Immunosorbent Assay for Rapid Determination of Sildenafil in Adulterated Functional Foods." *Food and Agricultural Immunology, 23*(4): 338–351.

Sonnen, Joshua A., Kathleen S. Montine, Joseph F. Quinn, Jeffrey A. Kaye, John C. S. Breitner, and Thomas J. Montine. (2008). Biomarkers for Cognitive Impairment and Dementia in Elderly People. *The Lancet Neurology,* 7(8): 704–714. Accessed July 10, 2021. doi.org/10.1016/S1474-4422(08)70162-5.

Stiksrud, Birgitte, et al. (2015). "Reduced Levels of D-dimer and Changes in Gut Microbiota Composition After Probiotic Intervention in HIV-Infected Individuals on Stable ART. *Journal of Acquired Immune Deficiency Syndrome,* 70(4): 329–337.

Syngai, Gareth Gordon, Ragupathi Gopi, Rupjyoti Bharali, Sudip Dey, G. M. Alagu Lakshmanan, and Giasuddin Ahmed. (2016). "Probiotics - the versatile functional food ingredients." *Journal of Food Science and Technology*, 53(2): 921–933.

Takezaki, Toshiro, Manami Inoue, Hiroki Kataoka, Syuhei Ikeda, Miyako Yoshida, Yoko Ohashi, Kazuo Tajima, and Suketami Tominaga. 2003. "Diet and Lung Cancer Risk From a 14-Year Population-Based Prospective Study in Japan: With Special Reference to Fish Consumption." *Nutrition and Cancer,* no. 2 (March): 160–67. https://doi.org/10.1207/s15327914nc4502_04.

Tecson-Mendoza, Evelyn Mae. 2007. "Development of Functional Foods in the Philippines." *Food Science and Technology Research,* no. 3: 179–86. https://doi.org/10.3136/fstr.13.179.

Teigler, Jeffrey E., Louise Leyre, Nicolas Chomont, Bonnie Slike, Ningbo Jian, Michael A. Eller, Nittaya Phanuphak, et al. 2018. "Distinct Biomarker Signatures in HIV Acute Infection Associate with Viral Dynamics and Reservoir Size." *JCI Insight,* no. 10 (May). https://doi.org/10.1172/jci.insight.98420.

Tokuda, Takahiko, Sultan A. Salem, David Allsop, Toshiki Mizuno, Masanori Nakagawa, Mohamed M. Qureshi, Joseph J. Locascio, Michael G. Schlossmacher, and Omar M. A. El-Agnaf. 2006. "Decreased α-Synuclein in Cerebrospinal Fluid of Aged Individuals and Subjects with Parkinson's Disease." *Biochemical and*

Biophysical Research Communications, no. 1 (October): 162–66. https://doi.org/10.1016/j.bbrc.2006.08.024.

Tran, Denise. 2013. "Development of an Enzyme-Linked Immunosorbent Assay (ELISA) for the Detection of Pecan Residues in Processed Foods." *Dissertations, Theses, and Student Research in Food Science and Technology,* August.

Trickey, Adam, Margaret T May, JorgJanne Vehreschild, Niels Obel, M John Gill, Heidi M Crane, Christoph Boesecke, et al. 2017. "Survival of HIV-Positive Patients Starting Antiretroviral Therapy between 1996 and 2013: A Collaborative Analysis of Cohort Studies." *The Lancet HIV,* no. 8 (August): e349–56. https://doi.org/10.1016/s2352-3018(17)30066-8.

Tsai, Yueh-Ting, Po-Ching Cheng, and Tzu-Ming Pan. 2012. "The Immunomodulatory Effects of Lactic Acid Bacteria for Improving Immune Functions and Benefits." *Applied Microbiology and Biotechnology,* no. 4 (September): 853–62. https://doi.org/10.1007/s00253-012- 4407-3.

Tumampo, Cherry Joy, Rio May Llanes, Arvie Joy Buncaras, Sandra Ganzo, Eleanor Beatriz Ragasa, Ernesto Balolong, Marilen Balolong, and Leslie Michelle Dalmacio. 2016. "Effect of *Lactobacillus plantarum* BS25 on cholesterol levels in vitro and in BALB/C mice (*Mus musculus* L.)." *Philippine Journal of Health Research and Development.*

Turk, Gabriela, Yanina Ghiglione, Macarena Hormanstorfer, Natalia Laufer, Romina Coloccini, Jimena Salido, César Trifone, et al. 2018. "Biomarkers of Progression after HIV Acute/Early Infection: Nothing Compares to CD4+ T-Cell Count?" *Viruses,* no. 1 (January): 34. https://doi.org/10.3390/v10010034.

Vaidya, Sagar A., Christian Korner, Michael N. Sirignano, Molly Amero, Sue Bazner, Jenna Rychert, Todd M. Allen, Eric S. Rosenberg, Ronald J. Bosch, and Marcus Altfeld. 2014. "Tumor Necrosis Factor α Is Associated With Viral Control and Early Disease Progression in Patients With HIV Type 1 Infection." *The Journal of Infectious*

Diseases, no. 7 (March): 1042–46. https://doi.org/10.1093/infdis/jiu206.

Van Cutsem, Eric, Claus-Henning Köhne, Erika Hitre, Jerzy Zaluski, Chung-Rong Chang Chien, Anatoly Makhson, Geert D'Haens, et al. 2009. "Cetuximab and Chemotherapy as Initial Treatmentfor Metastatic Colorectal Cancer." *New England Journal of Medicine,* no. 14 (April): 1408–17. https://doi.org/10.1056/nejmoa0805019.

Van Gils, Martijn P. M. Q., Erik B. Cornel, Daphne Hessels, W. Pim Peelen, J. Alfred Witjes, Peter F. A. Mulders, Harry G. Rittenhouse, and Jack A. Schalken. 2007. "Molecular PCA3 Diagnostics on Prostatic Fluid." *The Prostate,* no. 8: 881–87. https://doi.org/10.1002/pros.20564.

Van Nagell, J. R., P. D. DePriest, M. B. Reedy, H. H. Gallion, F. R. Ueland, E. J. Pavlik, and R. J. Kryscio. 2000. "The Efficacy of Transvaginal Sonographic Screening in Asymptomatic Women at Risk for Ovarian Cancer." *Gynecologic Oncology,* no. 3 (June): 350–56. https://doi.org/10.1006/gyno.2000.5816.

Van Tienen, A., Y. Hullegie, R. Hummelen, J. Hemsworth, J. Changalucha, and G. Reid. 2011. "Development of a Locally Sustainable Functional Food for People Living with HIV in Sub-Saharan Africa: Laboratory Testing and Sensory Evaluation." *Beneficial Microbes,* no. 3 (September): 193–98. https://doi.org/10.3920/bm2011.0024.

Van Veelen, W., J. W. B. de Groot, D. S. Acton, R. M. W. Hofstra, J. W. M. Höppener, T. P. Links, and C. J. M. Lips. 2009. "Medullary Thyroid Carcinoma and Biomarkers: Past, Present and Future." *Journal of Internal Medicine,* no. 1 (July): 126–40. https://doi.org/10.1111/j.1365-2796.2009.02106.x.

Van Weemen, B. K., and A. H. W. M. Schuurs. 1971. "Immunoassay Using Antigen-Enzyme Conjugates." *FEBS Letters,* no. 3 (June): 232–36. https://doi.org/10.1016/0014-5793(71)80319- 8.

Vanderstichele, Hugo, Karen De Vreese, Kaj Blennow, Niels Andreasen, Christian Sindic, Adrian Ivanoiu, Harald Hampel, et al. 2006. "Analytical Performance and Clinical Utility ofthe INNOTEST®

PHOSPHO-TAU(181P) Assay for Discrimination between Alzheimer's Disease and Dementia with Lewy Bodies." *Clinical Chemistry and Laboratory Medicine (CCLM),* no. 12 (January). https://doi.org/10.1515/cclm.2006.258.

Varadan, Vinay, Sitharthan Kamalakaran, Hannah Gilmore, Nilanjana Banerjee, Angel Janevski, Kristy L. S. Miskimen, Nicole Williams, et al. 2015. "Brief-Exposure to Preoperative Bevacizumab Reveals a TGF-β Signature Predictive of Response in HER2-Negative Breast Cancers." *International Journal of Cancer,* no. 3 (August): 747–57. https://doi.org/10.1002/ijc.29808.

Velazquez-Salinas, Lauro, Antonio Verdugo-Rodriguez, Luis L. Rodriguez, and Manuel V. Borca. 2019. "The Role of Interleukin 6 During Viral Infections." *Frontiers in Microbiology,* May. https://doi.org/10.3389/fmicb.2019.01057.

Venkataramana, M., R. Rashmi, Siva R. Uppalapati, S. Chandranayaka, K. Balakrishna, M. Radhika, Vijai K. Gupta, and H. V. Batra. 2015. "Development of Sandwich Dot-ELISA for Specific Detection of Ochratoxin A and Its Application on to Contaminated Cereal Grains Originating from India." *Frontiers in Microbiology,* May. https://doi.org/10.3389/fmicb.2015.00511.

Verhagen, Hans, Stefan Coolen, Guus Duchateau, Mark Hamer, Janet Kyle, and Andreas Rechner. 2004. "Assessment of the Efficacy of Functional Food Ingredients—Introducing the Concept 'Kinetics of Biomarkers.'" *Mutation Research/Fundamental and Molecular Mechanisms of Mutagenesis,* no. 1–2 (July): 65–78. https://doi.org/10.1016/j.mrfmmm.2004.01.013.

Vickers, Andrew J., and Simon F. Brewster. 2012. "PSA Velocity and Doubling Time in Diagnosis and Prognosis of Prostate Cancer." *British Journal of Medical and Surgical Urology,* no. 4 (July): 162–68. https://doi.org/10.1016/j.bjmsu.2011.08.006.

Villaño, D., A. Gironés-Vilapana, C. García-Viguera, and D. A. Moreno. 2016. "Development of Functional Foods." In *Innovation Strategies in the Food Industry,* 191–210. Elsevier. http://dx.doi.org/10.1016/b978-0-12-803751-5.00010-6.

Von Kleist, Sabine. 1986. "The Clinical Value of the Tumor Markers Ca 19/9 and Carcinoembryonic Antigen (Cea) in Colorectal Carcinomas: A Critical Comparison." *The International Journal of Biological Markers,* no. 1 (January): 3–8. https://doi.org/10.1177/172460088600100102.

Vona, Rosa, Lucrezia Gambardella, Camilla Cittadini, Elisabetta Straface, and Donatella Pietraforte. 2019. "Biomarkers of Oxidative Stress in Metabolic Syndrome and Associated Diseases." *Oxidative Medicine and Cellular Longevity,* May, 1–19. https://doi.org/10.1155/2019/8267234.

Vos, Alinda G., Annelieke Hulzebosch, Diederick E. Grobbee, Roos E. Barth, and Kerstin Klipstein-Grobusch. 2017. "Association between Immune Markers and Surrogate Markers of Cardiovascular Disease in HIV Positive Patients: A Systematic Review." Edited by Carmine Pizzi. *PLOS ONE,* no. 1 (January): e0169986. https://doi.org/10.1371/journal.pone.0169986.

Wadsworth, Jonathan DF, Andrew F Hill, Jonathan A Beck, and John Collinge. 2003. "Molecular and Clinical Classification of Human Prion Disease." *British Medical Bulletin,* no. 1 (June): 241–54. https://doi.org/10.1093/bmb/66.1.241.

Wallin, Anders, and Magnus Sjögren. 2001. "Cerebrospinal Fluid Cytoskeleton Proteins in Patients with Subcortical White Matter Dementia." *Mechanisms of Ageing and Development,* no. 16 (November): 1937–49. https://doi.org/10.1016/s0047-6374(01)00306-2.

Wang, Xi-Ying, Fen Zhang, Chi Zhang, Liang-Rong Zheng, and Jian Yang. 2020. "The Biomarkers for Acute Myocardial Infarction and Heart Failure." *BioMed Research International,* January, 1–14. https://doi.org/10.1155/2020/2018035.

Wu, Yunqing, Lifeng Zeng, Ying Xiong, Yuankui Leng, Hui Wang, and Yonghua Xiong. 2018. "Fluorescence ELISA Based on Glucose Oxidase-Mediated Fluorescence Quenching of Quantum Dots for Highly Sensitive Detection of Hepatitis B." *Talanta,* May, 258–64. https://doi.org/10.1016/j.talanta.2018.01.026.

Xiao, Yuhong, and Stuart N. Isaacs. 2012. "Enzyme-Linked Immunosorbent Assay (ELISA) and Blocking with Bovine Serum Albumin (BSA)—Not All BSAs Are Alike." *Journal of Immunological Methods,* no. 1–2 (October): 148–51. https://doi.org/10.1016/j.jim.2012.06.009.

Xiong, Xiangfeng, Dan Chen, Chaoyin Chen, Feng Ge, Diqiu Liu, Benyong Han, and Juan Fang. 2015. "Quantitative Indirect ELISA for Determination of Walnut Proteins in Foods." *Food Science and Biotechnology,* no. 6 (December): 2251–55. https://doi.org/10.1007/s10068-015-0300-z.

Xu, Changjiang, Christina Yong-Tao Li, and Ah-Ng Tony Kong. 2005. "Induction of Phase I, II and III Drug Metabolism/Transport by Xenobiotics." *Archives of Pharmacal Research,* no. 3 (March): 249–68. https://doi.org/10.1007/bf02977789.

Xu, Chuanlai, Liqiang Liu, Gang Cui, Xiaoling Wu, Hua Kuang, and Yuxi Chen. 2019. "Development of an Immunochromatographic Strip for the Detection of Rosiglitazone in Functional Foods Based on Monoclonal Antibodies." *Analytical Methods,* no. 38: 4910–16. https://doi.org/10.1039/c9ay01534e.

Xu, Liguang, Yuehong Song, Liqiang Liu, Shanshan Song, Jianping Zhu, Hua Kuang, and Chuanlai Xu. 2016. "Sandwich ELISA and Immunochromatographic Strip of Kunitz Trypsin Inhibitor Using Sensitive Monoclonal Antibodies." *Food and Agricultural Immunology,* no. 6 (March): 772–82. https://doi.org/10.1080/09540105.2016.1160367.

Yang, Shixin, Lucian Barbu-Tudoran, Marek Orzechowski, Roger Craig, John Trinick, Howard White, and William Lehman. 2014. "Three-Dimensional Organization of Troponin on Cardiac Muscle Thin Filaments in the Relaxed State." *Biophysical Journal,* no. 4 (February): 855–64. https://doi.org/10.1016/j.bpj.2014.01.007.

Younis, Kaiser, Saghir Ahmad, and Kausar Jahan. 2015. "Health Benefits and Application of Prebiotics in Foods." *Journal of Food Processing & Technology,* no. 04. https://doi.org/10.4172/2157-7110.1000433.

Zhan, Jiaxin, Shanshan Qin, Lili Lu, Xiamin Hu, Jun Zhou, Yeying Sun, Jian Yang, et al. 2016. "MiR-34a Is a Common Link in Both HIV- and Antiretroviral Therapy-Induced Vascular Aging." *Aging,* no. 12 (November): 3298–3310. https://doi.org/10.18632/aging.101118.

Zhang, Jie, Zhewen Chen, Yanwen Lu, Daoyuan Tu, Fengqian Zou, Shouwen Lin, Weinan Yu, Mingyong Miao, and Hanping Shi. 2021. "A Functional Food Inhibits Azoxymethane/Dextran Sulfate Sodium-Induced Inflammatory Colorectal Cancer in Mice." *OncoTargets and Therapy,* February, 1465–77. https://doi.org/10.2147/ott.s283465.

Zhang, Yan, Tao Pan, Guozhen Fang, Dongyue Ma, and Shuo Wang. 2009. "Development of a Solid-Phase Extraction-EnzymeLinked Immunosorbent Assay for the Determination of 17beta-19-Nortestosterone Levels in Antifatigue Functional Foods." *Journal of Food Science,* October. https://doi.org/10.1111/j.1750-3841.2009.01311.x.

Zhou, Shan-shan, Jing-peng Jin, Ji-qun Wang, Zhi-guo Zhang, Jonathan H Freedman, Yang Zheng, and Lu Cai. 2018. "MiRNAS in Cardiovascular Diseases: Potential Biomarkers, Therapeutic Targets and Challenges." *Acta Pharmacologica Sinica*, no. 7 (June): 1073–84. https://doi.org/10.1038/aps.2018.30.

Zhu, Longjiao, Jing He, Xiaohan Cao, Kunlun Huang, Yunbo Luo, and Wentao Xu. 2016. "Development of a Double-Antibody Sandwich ELISA for Rapid Detection of Bacillus Cereus in Food." *Scientific Reports,* no. 1 (March). https://doi.org/10.1038/srep16092.

In: ELISA
Editor: Jose Ma. M. Angeles
ISBN: 978-1-68507-500-2

Chapter 6

A Refined Rapid Antibody-Based Immunoassay Improved Sensitivity of Paralytic Shellfish Poisoning Toxins Detection

***Chun-Kwan Wong*[*], *Pui-Yi Cheung*[†] *and Rickjason Chi-Wai Chan*[‡]**
Public Health Laboratory Services Branch,
Centre for Health Protection, Department of Health,
Hong Kong, China

Abstract

Paralytic shellfish poisoning (PSP) is a significant public health concern that causes foodborne illness in humans through the consumption of filter-feeding shellfish tainted with PSP-toxins. Although PSP mouse bioassay (MBA) is an official method of the

[*] Corresponding Authors' E-mail: so_phls5@dh.gov.hk.
[†] Corresponding Authors' E-mail: so_phls4@dh.gov.hk
[‡] Corresponding Authors' E-mail: rickjasonchan@dh.gov.hk.

AOAC International and a reference method of European Union legislation for the quantitative examination of PSP toxicity in shellfish, its main drawbacks include poor sensitivity, undesirable detection limit and issues of animal ethics. In an attempt to reduce the use of laboratory mouse bioassay for PSP toxicity test, Enzyme-Linked Immunosorbent Assay (ELISA) is one of the desirable means for providing a fast and highly sensitive antibody-based test to screen out potential PSP-toxins contamination in shellfish samples. Working on the same principle as the ELISA, Lateral Flow Immunosorbent Assay (LFIA) is beneficial to providing a one-step rapid test with sensitivity superior to the traditional animal assay. However, the low cross-reactivity of immunological tests to gonyautoxins 1,4 (GTX1,4) was still a pivotal concern in the past decades. One of the commercially available LFIAs had the test kits and procedure refined to include an additional chemical hydrolysis step to transform GTX1,4 into neosaxitoxin (NEO), a carbamate toxin with higher cross-reactivity. The chemical treatment, at the same time, converts gonyautoxins 2,3 (GTX2,3) into saxitoxin (STX) with the highest cross-reactivity. The present study verified that the refined method was effective for improving the sensitivity of the assay for common types of shellfish matrices containing GTX1,4 at a regulatory level. The sensitivities to other PSP-toxins (e.g., NEO) in different acidic shellfish extracts were also enhanced after implementing the hydrolysis reaction. As for samples spiked with PSP-toxins, a retrospective examination of PSP-toxins contaminated field samples also demonstrated the same efficacy in sensitivity tests and improved performance for PSP-toxins detection when the refined procedures were applied.

Keywords: gonyautoxins, lateral flow immunochromatographic assay, neosaxitoxins, paralytic shellfish poisoning, rapid kit, shellfish toxins

INTRODUCTION

Paralytic shellfish poisoning (PSP) is a well-known foodborne intoxication through the consumption of filter-feeding molluscan shellfish naturally contaminated with high levels of PSP-toxins. Marine dinoflagellates (Dinophyceae) such as *Alexandrium*, *Gymnodinium,* and *Pyrodinium* are microalgae that produce non-protein neurotoxins which could be transferred directly through the food chain ending up to humans

(Lehane, 2000). To date, more than 55 PSP-toxin analogs were discovered (Wiese et al., 2010). In general, the toxins can be classified into three major structural sub-groups in descending order of toxicity (Luckas, Hummert and Oshima, 2003): (1) carbamate toxins (saxitoxin (STX), neosaxitoxin (NEO), and gonyautoxins 1–4 (GTX1–4)), (2) decarbamate toxins (e.g., dcSTX, dcNEO, and dcGTX1–4), and (3) N-sulfocarbamate toxins (e.g., gonyautoxins 5–6 (GTX5–6 or B1–2) and C-toxins (C1–4)).

STX (guanidine compound), one of the Schedule 1 chemical warfare agents in the list of the Chemical Weapons Convention by Organization for the Prohibition of Chemical Weapons (http://www.opcw.org), is the most potent PSP-toxin with oral and inhalation LD_{50} in humans at 5.7 μg/kg (Dorner et al., 2016) and 5 mg·min/m^3, respectively (Pitschmann and Hon, 2016). Its lethal dose in humans is estimated to be as low as 0.3 mg (Gad, 2014). It is also used as a toxicity reference to other PSP-toxins wherein the relative toxicities of C-toxins have the least effects in comparison with the others (Oshima, 1995a, b; Rourke et al., 2008; EFSA, 2009). The PSP-toxins can block the voltage-gated sodium channels of nerve cells, causing inhibition of the normal propagation of nerve impulses. The clinical symptoms of PSP-toxins in humans include mainly neurological manifestations varying from slight numbness/tingling sensations, muscular weakness to fatal respiratory paralysis, or even death while gastrointestinal disorders are less common (Wong et al., 2010, 2011). The symptoms can be manifested within an hour after poisoning.

Many food-regulatory authorities have put considerable efforts to monitor these marine biotoxins in edible shellfish in the market. The reference method recommended by the European Union (EU) for determining the level of PSP-toxins in shellfish tissue is the mouse bioassay (MBA) (Commission regulation EC 2042/2005), with which the regulatory action limit for PSP toxins in shellfish is set at 80 μg saxitoxins equivalents (μg STXeq) per 100 grams of shellfish tissue (EC Regulation No. 853/2004) (Anons, 2005). In Hong Kong, it is stipulated in the Public Health and Municipal Services Ordinance that all food available for sale on the market must be fit for human consumption. The use of non-animal assay for monitoring is an ethical option in the field against the utilization of live experimental mice for toxicity analysis. In order to encourage more humane alternatives, the "Three Rs" (reduction, replacement and refinement) requirements on the protection of animals in bioassay and scientific experiments have been stipulated in Directive 86/609/EEC and Directive 2010/63/EU in the past decades.

The Scotia Rapid PSP test (SRPT) is a commercially available antibody-based immunoassay manufactured by Scotia Rapid Testing Ltd (Nova Scotia, Canada), formerly known as Jellett Rapid Testing Ltd and MIST Alert™, for simple and rapid PSP-toxins screening in shellfish samples (Jellett et al., 2002; Mackintosh et al., 2002; Turner et al., 2015). Working on the same principle as enzyme-linked immunosorbent assays (ELISA) commonly used in many medical diagnostics, SRPT is a lateral flow immunochromatographic strip-format test that adopts polyclonal PSP-toxin antibodies for the qualitative detection of the toxins. The basic configuration of the SRPT strip is shown in Figure 1.

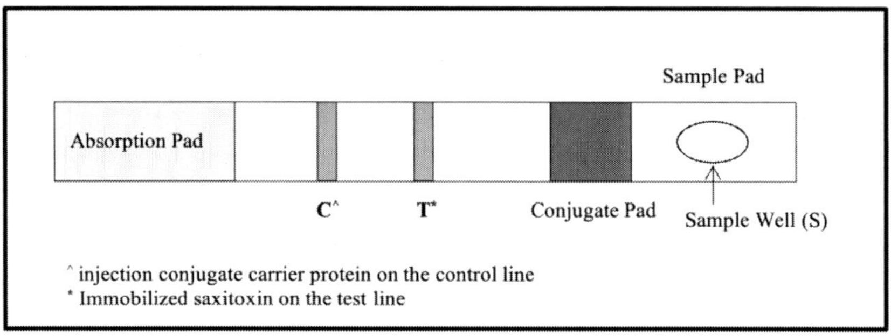

Figure 1. Lateral flow immunochromatographic test strip of Scotia Rapid PSP test (SRPT) (Laycock, Donovan and Easy, 2010).

The PSP-toxins in the sample bind competitively with bound STX on the test strip (i.e., the immobilized STX on the test line) against colloidal gold-labeled anti-STX antibodies. The higher the concentration of toxins added, the less intense the color of the T-line on the strip will develop. The color intensity of the C-line (control line) on the strip should be intense enough for assigning the test result as a valid test, indicating that antigen/antibody reaction and liquid flow on the strip have not interfered. The average detection limit of the kit is around 40 μg STXeq per 100 g of shellfish tissue as stated by the manufacturer.

The assay was evaluated previously to detect PSP toxins at levels close to the Limit of Detection (LOD) of the PSP MBA (AOAC 959.08), which ranges between 20 and 70 μg STXeq per 100 g shellfish tissue in some studies (Laycock, Donovan and Easy, 2010; Wong et al., 2010; Turner et al., 2015; Dorantes-Aranda et al., 2017). In an attempt to reduce the use of laboratory animals in bioassays for toxicity testing, this type of antibody-based assay is one of the desirable alternatives for providing a fast and simple means to screen out potential PSP-toxins contamination in shellfish samples under routine monitoring program (Usleber et al., 2001).

In the PSP-toxins surveillance program of our laboratory, only SRPT positive/suspected samples or invalid results would require a confirmatory test by MBA. It should be noted, however, that the low cross-reactivity of this immunological test to GTX1,4 is still a major concern (Cefas, 2006; Turner et al., 2014; Harrison, Johnson and Turner, 2016; Dorantes-Aranda et al., 2017; Dorantes-Aranda et al., 2020). Nevertheless, the assay has been refined by the manufacturer to include an additional chemical hydrolysis step to convert GTX1,4 into NEO, one of the carbamate toxins with higher cross-reactivity. The same chemical treatment also transforms GTX2,3 into STX with the highest cross-reactivity. As a result, the sensitivity of the assay for samples containing high proportions of GTX1,4 can be increased from 1.8–3.3% to 26%. This refined method provided greater detection sensitivity to GTX1,4 in some acidic (1% acetic acid) sample extracts containing high proportions of these epimers (Turner et al., 2015; Harrison, Johnson, and Turner, 2016; Dorantes-Aranda et al., 2017).

The objective of the present work is to verify the performance and applicability of the refined SRPT method for improving the sensitivity of the assay for common types of shellfish matrices potentially containing GTX1,4 (epimer pair) at the regulatory level. The sensitivities of the assay to other common PSP-toxins were also studied. Certified PSP toxin standards were used to spike shellfish hydrochloric acid (0.1–0.18N) extracts. Field shellfish samples naturally contaminated with PSP-toxins were also used for a retrospective investigation and evaluation of the efficacy and performance of the refined method on PSP-toxins detection.

Materials and Methods

Shellfish Collection, Samples Extraction, and Preparation

Common raw shellfish samples were obtained from markets in Hong Kong under the government routine PSP-toxins surveillance program. All samples were kept frozen and thawed at 4°C before toxin extraction in accordance with the method as described in Wong et al. (2010). Briefly, 100-150 grams of shellfish tissues were used for the extraction of toxins. Acidic (0.1–0.18N diluted hydrochloric acid) extracts of shellfish including oysters, scallops (raw scallops with shells), scallops (M) (scallops adductor muscle only), geoducks, and clams previously tested negative (i.e., below detection limit of the assay) for PSP-toxins by SRPT were used for the toxins spiking test. PSP-toxin Quality Control (QC) scallop extract (equivalent to 80 μg STX/100 g of shellfish tissue) was used as a positive control. Five SRPT negative sample extracts (oysters, scallops, scallops (M), geoduck, and clams) were selected as negative controls. The acidic extracts of the shellfish samples were stored at -20°C and then kept at 4°C for a day before the PSP-toxins spiking experiment.

PSP-Toxins Standards

Certified toxin standards, STX, NEO, GTX1–4, dcSTX, dcGTX2–3 and GTX5, were purchased from the Certified Reference Materials Program of the Institute for Marine Biosciences, National Research

Council (NRC), Halifax, Nova Scotia, Canada, for measuring the sensitivity of SRPT to regulatory action limit of PSP-toxins in different shellfish matrices. All standard solutions were stored at 4°C, with their expiry dates and certificates were checked before use.

Evaluation of the Sensitivity of Existing and Refined SRPT Methods to PSP-Toxins Standards in Different Spiked Shellfish Matrix Media

The matrix effects of some common shellfish extracts on SRPT performance were evaluated with PSP-toxins standards. Shellfish samples previously determined as PSP-toxins negative by SRPT were selected. Acidic extracts of these shellfish samples were prepared and used as the matrix media. In general, certified PSP-toxins standards were spiked individually to different acidic shellfish extracts at or above the regulatory level (80 μg STXeq/100 g of shellfish tissue) to evaluate the sensitivity performance of the SRPT kit before (i.e., the existing method (EM) as described in Wong et al., 2010) and after the application of the refined method (RM).

Expiry dates and Certificates of Analysis of the kits and kit reagents were checked before use. The manufacturer's instructions for the kits were strictly followed. Briefly, 100 μl of the toxin spiked shellfish extract was mixed with 400 μl of the kit running buffer by pipetting up and down at least three times. 100 μl of the resulting mixture was inoculated into the sample well "S" on an SRPT strip which was then analyzed between 35 and 60 minutes. A positive control (0.4 μg/ml STX dihydrochloride of NRC of Canada in 0.003N hydrochloric acid) was included in each run to ensure the sensitivity and efficacy of the test.

The result interpretation was based on the criteria of color intensity of the C- and T-line on the SRPT strip as stated in the Test Interpretation Sheet provided with each PSP test kit. For a negative result (-ve), the T-line is darker than the reference positive T-line on the Test Interpretation Sheet. For an invalid result, the intensity of the C-line is equal to or fainter than that of an invalid example. For a PSP-toxins positive result, the intensity of the T-line is equal to or fainter than the reference positive T-line. The positive results were rated in three levels as described in Wong et al. (2010): strongly +ve, +ve (the ratio of color intensity between C-line and T-line being approximately equivalent to that of the positive result shown on the Test Interpretation Sheet), and slightly +ve. Interpretation remarks could be included to record the details of the color intensity of both C- and T-line. The test results were further verified by another trained public health laboratory analyst to ensure that a consensus was reached on the interpretation before confirming the final results.

For the RM procedures, 200 μl of the toxin spiked shellfish extract was first added to the reagent powder in a clear vial provided by the SRPT manufacturer. The mixture was pipetted up and down at least three times and was allowed to incubate at room temperature for 60 minutes. 100 μl of the extract-reagent powder mixture was then mixed with 400 μl of the kit running buffer by pipetting up and down at least three times. The sample well "S" on an SRPT strip was then inoculated with 100 μl of the resulting mixture and the result was read between 35 and 60 minutes.

Field and Clinically-Linked Samples (FS and CS)

Acidic extract of some PSP-toxins positive samples including eight FS (one crab sample was used) and three CS acidic extract shellfish samples were selected to test the efficacy of the refined method for PSP-toxins detection in a retrospective study. The results were compared with those of the existing method. These samples were previously tested by the official AOAC mouse bioassay procedures for paralytic shellfish poison as described in AOAC (2000) and Wong et al. (2010). Seven samples, including scallops and geoduck clams, were also tested positive for PSP-toxins by using the high-performance liquid chromatography (HPLC) technique (Oshima, 1995).

Statistical Analysis

Sensitivities of SRPT (existing and refined methods) to individual PSP-toxins in different shellfish matrices were compared using the following algorithm and equations following the method as stated in ISO 16140-2 (2016):

EM positive (+) and RM positive (+) = +/+ Positive Agreement [PA]

EM negative (-) and RM negative (-) = -/- Negative Agreement [NA]

EM positive (+) and RM negative (-) = +/- Negative Deviation [ND]

EM negative (-) and RM positive (+) = -/+ Positive Deviation [PD]

(Invalid results obtained in the existing method were considered as negatives in the calculation)

Sensitivity of refined method (SE_{RM}) = $[(PA + PD)/(PA + ND + PD)] \times 100\%$ (1)

Sensitivity of existing method (SE_{EM}) = $[(PA + ND)/(PA + ND + PD)] \times 100\%$ (2)

Chi-square test (SigmaPlot® 12.0, SPSS Inc.) and examination of discordant results or disagreements (i.e., PD and ND) (ISO 16140:2003E, 2003) were conducted to compare the results of the two methods.

Acceptance Criteria

The refined PSP rapid detection method would be accepted and applied to our routine PSP-toxins surveillance program if the test results obtained were non-inferior to those obtained by the existing PSP rapid detection method. The RM would not be accepted if any negative deviation (false negative) result was obtained for samples spiked with toxins at or above the regulatory level.

RESULTS

The performance of SRPT on the detection of various PSP-toxins spiked to different shellfish matrices is presented in Table 1. In general, PSP-toxin standards, except GTX1,4, spiked to shellfish samples at the regulatory level (80 μg STXeq/100 g of shellfish tissue) could be detected by the kit using the existing method (EM). It should be noted, however, that some PSP-toxin standards such as NEO, dcGTX2,3 could not be detected by the kit at the regulatory level in some matrices of clams, scallops, and scallops (M). Considerable attention should be placed on oyster samples spiked with those PSP-toxin standards. Invalid results were produced by the kit at the regulatory limit of all tested PSP-toxins or even at a level 8 times higher than the limit.

On the contrary, the application of the refined method (RM) could significantly improve the results obtained by the EM. Overall, the negative results obtained from NEO, dcGTX2,3, and GTX1,4 spiked shellfish extracts (clams and scallop (M)) were subsequently tested positive or even strong positive after allowing the hydrolysis of the acidic extracts with the reagent powder for 60 minutes. Moreover, for all those invalid results obtained in oyster matrix media, it was noted that positive or strong positive results could be achieved after the application of the hydrolysis step in the RM. The pink color of most C- and T-lines on the strips were more intense and easier to read than those without treatment with the reagent powder.

The performance of SRPT on PSP-toxins detection in the field and clinically-linked samples is shown in Table 2.

Table 1. The performance of SRPT on the detection of various PSP-toxins in different shellfish matrix media*

PSP-toxins	μg/100 g shellfish	Existing method result ( ): repeated test result	Refined method result	Existing method result ( ): repeated test result	Refined method result	Existing method result ( ): repeated test result	Refined method result
		Oysters matrix media		*Scallops matrix media*		*Geoducks matrix media*	
STX	80	Invalid [a] (Invalid [a])	Strongly +ve [b]	Strongly +ve [c]	Strongly +ve [b]	Strongly +ve [c]	Strongly +ve [b]
dcSTX	80	Invalid [a]	Strongly +ve [c]	+ve	Strongly +ve [c]	+ve	Strongly +ve [c]
NEO	80	Invalid [a] (Invalid [d])	+ve	Invalid (+ve)	Strongly +ve [c]	+ve	Strongly +ve [c]
GTX2,3	80	Invalid [a,e]	Strongly +ve [b]	Strongly +ve [c]	Strongly +ve [b]	Strongly +ve [c]	Strongly +ve [b]
GTX5	40	-	-	-	-	+ve	Strongly +ve [c]
GTX5	80	Invalid [a]	Strongly +ve [c]	+ve	Strongly +ve [b]	-	-
dcGTX2,3	80	Invalid [a,e]	Strongly +ve [b]	+ve	Strongly +ve [c]	+ve	Strongly +ve [b]
GTX1,4	80	Invalid [a]	+ve	-ve	+ve	-ve	+ve
GTX1,4	160	Invalid [a]	+ve	+ve [f]	+ve	-ve	+ve
GTX1,4	320	Invalid [a]	+ve	-	-	-	-
GTX1,4	640	Invalid [a]	Strongly +ve [c]	-	-	-	-
		Clams matrix media		*Scallops (M) matrix media*		*Positive control shellfish extracts*	
STX	80	Strongly +ve [c]	Strongly +ve [b]	Strongly +ve [c]	Strongly +ve [b]	STX spiked scallops, 80 μg/100 g shellfish	
dcSTX	80	Strongly +ve [c]	Strongly +ve [c]	+ve	Strongly +ve [c]	Strongly +ve [b]	Strongly +ve [b]
NEO	80	-ve [g]	Strongly +ve [c]	-ve [g]	+ve		
GTX2,3	80	Strongly +ve [c]	Strongly +ve [b]	Strongly +ve [c]	Strongly +ve [b]		
GTX5	80	-	-	-	-	*Negative control shellfish extracts*	
dcGTX2,3	80	-ve [g]	Strongly +ve [c]	+ve	Strongly +ve [c]	Oysters Scallops Geoducks Clams Scallops (M)	
GTX1,4	80	-ve	+ve	-ve [g]	+ve	-ve	-ve
GTX1,4	160	-ve [h]	+ve	-ve [h]	+ve		

*Results highlighted with gray are invalid/negative; [a]The intensity of C-line was equal to that of invalid example illustrated on the instruction sheet; [b]No visible T-line was seen except C-line; [c]Shadow on T-line; [d]The intensity of C-line was fainter than that of invalid example illustrated on the instruction sheet; [e] T-line was fainter than that of invalid example illustrated on the instruction sheet; [f]T-line was fainter than that of +ve T-line illustrated on the instruction sheet; [g] C-line was darker than T-line; [h] the color intensity of C- and T-lines were the same

Table 2. Comparison of SRPT results from the existing and refined methods using field and clinically-linked samples

Year of collection	Acidic shellfish extract	Field sample (FS)/Clinically-linked sample (CS)	MBA result μg STXeq/100 g shellfish	Existing method result	Refined method result	Rating + (existing)	Rating + (refined)
2007	Scallop A	FS	1860^	Strongly +ve*	Strongly +ve*	1	1
2007	Scallop B	FS	2430^	Strongly +ve*	Strongly +ve*	1	1
2010	Scallop C	CS	40^	-ve/slightly +ve	+ve#	1	1
2010	Scallop D	CS	37^	+ve	+ve#	1	1
2014	Geoduck A	FS	549^	Strongly +ve*	Strongly +ve*	1	1
2014	Geoduck B	FS	50^	Strongly +ve*	Strongly +ve*	1	1
2015	Geoduck C	FS	44^	Slightly +ve*	Strongly +ve*	1	1
	No HPLC result as reference for the following samples						
2017	Bloody clam	FS	41	Strongly +ve *	Strongly +ve*	1	1
2019	Whelk A	CS	Negative	Invalid	Slightly +ve#	0	1
2019	Whelk B	FS	Negative	-ve	-ve/slightly +ve#	0	0/1
2019	Crab (not shellfish)	FS	Negative	Invalid	-ve	0	0

^PSP-toxins were detected in the sample by high-performance liquid chromatography (HPLC) analysis. For Scallop A and Whelk B, the -ve/slightly +ve result was regarded as suspicious/undetermined result and MBA was performed for confirmation.

* No visible T-line was seen except C-line

The color intensity of C- and T-lines were the same

+ For statistical analysis, 0 = invalid/-ve result; 1 = slightly positive/positive/strongly positive result (As no HPLC result was available for the last four samples Bloody clam, Whelk A, Whelk B and Crab, the ratings are not included in the statistical analysis)

Table 3. Sensitivities of existing (SE_{EM}) and refined (SE_{RM}) methods of SRPT to PSP-toxins in different shellfish matrices (spiked at or below 80 µg STXeq/100 g shellfish tissue and naturally contaminated shellfish at MBA toxicity levels between 37 and 2430 µg STXeq/100 g shellfish tissue)

Shellfish matrix	Spiked samples	Field & clinically-linked samples ^	SE_{EM} (%)	SE_{RM} (%)
Oyster (n = 7)	7	-	0	100
Scallops (n = 11)	7	4	45	100
Geoducks (n = 10)*	7	3	90	100
Clams (n = 6)	6	-	50	100
Scallop (M) (n = 6)	6	-	67	100
Total tested (n = 40)^			65	100

* Including one sample spiked with 40 µg STXeq/100 g of shellfish tissue (see Table 1: Geoducks).

^ Including 7 samples in the retrospective study with PSP-toxins detected in both mouse bioassay and high-performance liquid chromatography analysis.

Similar to the toxins spiking study, the adoption of an additional hydrolysis step to those PSP-toxins tainted field sample extracts had achieved the equivalent or better performance when compared with the original method without treatment with the reagent powder. For the two invalid results of Whelk A and Crab as determined by the EM, the application of the refinement step in the RM could subsequently produce valid results. A slightly positive result was demonstrated for the clinically-linked Whelk A sample, though it was negative in MBA.

Concerning the Whelk B sample with the negative result using the existing method, an inconclusive result (-ve or slightly +ve) was obtained after the hydrolysis step whilst the MBA result was negative. Although no HPLC results were available for further comparison, testing the four FS and CS samples (Bloody clam, Whelk A, Whelk B, and Crab) with the RM had eliminated two invalid results and revealed that the original invalid result of Whelk A and the negative result of Whelk B might be slightly positives missed by the EM and the MBA. As the detection limit of antibody-based SRPT is lower than that of MBA and its sensitivity is superior to animal-based test, the slightly positive results obtained for some samples with the RM might implicate the presence of trace levels of PSP-toxins, probably lower than 40 µg STXeq/100 g of shellfish tissue.

Sensitivities of the existing and the refined methods to PSP-toxins in different spiked shellfish matrices are shown in Table 3. The RM had a sensitivity of 100% on all shellfish extracts inoculated with PSP-toxins at and above the regulatory level (80 μg/100 g shellfish). The same level of sensitivity could also be achieved in the retrospective study on the seven field and clinically-linked shellfish samples (FS and CS) that were previously tested positive in both MBA and HPLC analysis. In addition, no false-positive result was found when they were tested retrospectively with the refined method as compared with the HPLC results. Overall, the application of the RM increased the sensitivity of SRPT to PSP-toxins from 0–90% to 100% in various shellfish matrices.

Statistical analysis of the results of the EM and RM on 40 samples of spiked or naturally contaminated shellfish matrices using the chi-square test produced a value of 25.827 (1 degree of freedom), indicating that the two methods were different at $P \leq 0.001$. Discordant results (i.e., PD and ND) were also examined using the equation Y = PD + ND (ISO 16140:2003E, 2003) where "Y" is the total number of discordant results or disagreements and was equal to 21 (i.e.,: Y = 21 + 0; PD is 21 and ND is 0; Table 1 & 2). Y was between 6 and 22 disagreements, and the value of "m", being the smallest of the two values of PD and ND, was zero. Using the binomial law to obtain the value of M for given Y disagreements, the value of M (for $\alpha < 0.05$) was then derived as 5. Since m was less than M at Y = 21, the two methods were different at $\alpha < 0.05$ (2-sided).

Conclusion and Discussion

Although paralytic shellfish poison mouse bioassay is an official method (Official Method of Analysis, OMA) of the AOAC International and a reference method of European Union legislation, the main drawbacks of this method include the potential of getting false positive and negative results due to extraction pH, metal contaminations (e.g., zinc) and "salt effect" in the samples (McFarren, 1959; Park, Guzman-Perez and Lopez-Garcia, 1986; LeDoux and Hall, 2000; Vale et al., 2008). In addition, the ethical issue of the animal assay is also of crucial importance in its laboratory application in routine PSP tests. Antibody-based analysis such as ELISA is one of the alternatives to replacing the animal-based toxicity test. As ELISA is suitable for use as a screening tool for some marine toxins due to its desirable sensitivity and selectivity for structural recognition of toxin analogs or derivatives (Reverte et al., 2014; Eangoor et al., 2017), several ELISA kits for PSP-toxins test were developed and used for field applications on both shellfish and clinical samples (DeGrasse et al., 2014; Kawatsu et al., 2014; Harrison, Johnson, and Turner, 2016; Dorantes-Aranda et al., 2017; Eangoor et al., 2017; Wharton et al., 2017). However, conventional ELISA tests require more laboratory equipment and techniques and, at the same time, involve complicated workflows and are more time-consuming when compared with the rapid one-step Lateral Flow Immunosorbent Assay (LFIA) tests. In the past few years, LFIA tests have been increasingly adopted in the clinical, food safety, and environmental analysis considering their simple test procedures, lower cost, high throughput, and shorter turnaround time (Posthuma-Trumpie, Korf and van Amerongen, 2009; Sajid, Kawde and Daud, 2015; Turner et al., 2020).

The Scotia Rapid PSP assay (formerly MIST Alert™) is one of the competitive LFIA tests that has been used for almost two decades for screening potential contamination of PSP-toxins in shellfish since its first launch by Jellett Biotek Ltd, Canada (Jellett et al., 2002; Mackintosh et al., 2002). However, its low cross-reactivity to both GTX1,4 and NEO as well as the sporadic occurrence of invalid test results have been known for years (Cefas, 2006; Dorantes-Aranda et al., 2017; Wong et al., 2010; Turner et al., 2014). The application of the refined method for acetic acid shellfish extracts has recently been proven effective to improve the sensitivity of the assay to shellfish samples in the UK containing high proportions of GTX1,4 (Turner et al., 2015). In this study, when hydrochloric acid shellfish extracts (AOAC Official Method 959.08) were tested with the application of the refined method, the sensitivity of the assay for samples containing N-1-hydroxylated toxins, GTX1,4, and NEO at the regulatory level could *de facto* be improved. The sensitivities to other PSP-toxins were also enhanced when compared with the original method. The chemical reagent powder crystals provided by the manufacturer contain a thiol reagent which may convert those low cross-reactivity GTX1,4 epimers to NEO with higher cross-reactivity (Laycock, Kralovec and Richards, 1995; Turner et al., 2015). This hydrolysis reaction, at the same time, can transform non-N-1-hydroxylated GTX2,3 to STX (Laycock, Kralovec and Richards, 1995). Such reaction involves the cleavage of sulfate from the 11-position of the GTX1,4 and GTX2,3 molecules.

In the retrospective study of field samples, even though slightly positive and undetermined results were found respectively in Whelks A and B after the treatment with the reagent powder, further toxicity confirmation tests by MBA were negative. It was most probably due to the detection limit of the mouse bioassay (about 40μg STXeq/100 g of shellfish tissue) (Park, Guzman-Perez and Lopez-Garcia, 1999) being inferior to that of the antibody-based assay (at or below 40 μg STXeq/100 g of shellfish tissue) (Jellett et al., 2002; Wong et al., 2010).

In general, test results obtained with the refined method were non-inferior and easier to read than those obtained from the original procedures without the chemical treatment by thiol reagent. According to our previous study (Wong et al., 2010), STX, NEO, GTX2,3, and GTX5 were the most commonly occurring toxins in the shellfish samples in Hong Kong. GTX2,3 and NEO were also commonly found in shellfish samples from southern China, as well as worldwide (Costa et al., 2009; García et al., 2010). As the additional chemical conversion step could convert GTX1,4 and GTX2,3 to congeners of NEO and STX, respectively, it is expected that due to higher cross-reactivity, more samples would become positive in the refined screening test and necessitate toxicity confirmation by MBA. This would cause a potential burden on the increased demand of MBA for those samples which could have been negative in the original SRPT without the refinement steps. However, from a public health perspective, the chance of missing those potentially tainted samples for further examination might be lower and the risk of paralytic shellfish poisoning would be reduced in the general public. On the contrary, such a potential burden might be reduced to a certain extent by applying the refined method decreasing the number of invalid results, especially for oyster samples, that would entail confirmatory testing by MBA.

For the operational application of the refined steps in routine laboratory services, the main disadvantages were the concerns of a relatively longer turnaround time wherein additional 60 minutes of incubation time was required to complete the hydrolysis reaction. Also, additional steps and consumables will be required for the test. Additional laboratory procedures were relatively simple to handle where expertise training and highly technical work were not required for the refinement steps. After considering the cost-effectiveness and ethical concerns on the use of the animal test, the refined method was considered more suitable to be used for re-testing of samples with indeterminate/invalid test results obtained from the existing method, for samples suspected to contain high levels of GTX or interfering substances, and for testing clinically linked samples. Additionally, the use of an electronic scanning device, Skannex system (Scotia Rapid Testing Ltd, Canada) with the SRPT kit software for automatic interpretation of the control and test lines on the SRPT cartridges was also recommended to eliminate the subjectivity of color interpretation by analysts. Such a tool could further enhance the efficiency of the assay by mass screening of shellfish samples for PSP-toxins contamination (Turner et al., 2015).

Overall, the refinement of SRPT with a hydrolysis step was effective to improve the sensitivity of the assay to the PSP-toxins tested. However, this assay, as in many other ELISA tests, does not provide information on toxin identification present in the samples. Further analysis on the samples is therefore needed to identify the toxins. Also, as there were no mussel and other less common shellfish samples available for testing in this study, the sensitivity and efficacy of the refined method on these shellfish matrices were still questionable. Further studies are required to elucidate the effect of other shellfish matrices or possible interferences on the performance of this refined PSP-toxin detection method. Komano, Maruko, and Sekiguchi (2011) published a study suggesting that this lateral flow immunoassay kit should be used for on-site detection of STX during counterterrorism activities. Considerable attention should also be placed on the findings that contamination with formaldehyde (0.088%) and sodium hydroxide (0.05 M) solutions could trigger false-positive results in the analysis of STX. False-negative results could also be possible in the presence of low levels of sodium hypochlorite (0.00002%, 1:249999). These results showed that the performance of the assay would be easily affected by some common chemicals even in trace quantities.

Scotia Rapid Testing Ltd, Canada closed its operations in early 2021 and the production and sales of SRPT kits were taken up by another company in Chile. The same situation related to the stability and continuity of product supply from manufacturers could also be encountered with other commercially available laboratory testing products. To gradually reduce and hopefully move away from the use of laboratory mouse bioassay, and to prevent the interruption of routine public health laboratory testing and monitoring of PSP-toxins in shellfish products, exploring other immunology-based alternatives and supplies as well as other analytical chemical means as backup tests are advantageous and undoubtedly necessary.

Disclaimer

All commercial products or trade names mentioned in this article do not represent any suggestion, approval, or endorsement for use by the authors' department and division. All laboratory mice used in the bioassay were sacrificed after experiments in compliance with the Code of Practice for Care and Use of Animals for Experimental Purposes, of Agriculture, Fisheries and Conservation Department and the guidelines of the Department of Health, Government of Hong Kong SAR, China.

Acknowledgments

The authors would like to express their gratitude to the staff of the Centre for Food Safety, Food and Environmental Hygiene Department, for the collection of shellfish samples for PSP-toxins examination in the routine surveillance program and all staff of Non-Clinical Microbiology Laboratory, Public Health Laboratory Services Branch, Centre for Health Protection, Department of Health, Government of Hong Kong Special Administrative Region who demonstrated their professionalism in public health services, especially during the COVID-19 anti-epidemic period in Hong Kong.

References

AOAC (2000). AOAC Official Method 959.08. Paralytic Shellfish Poison Biological Method. Final Action. *AOAC Official Methods of Analysis*, 17th edition. Gaithersburg, Maryland, 59-61.

Anon (2005). AOAC official method 959.08. Paralytic shellfish poison. Biological method. Final action. In: Truckses, M. W. (Ed.), *Natural Toxins. AOAC Official Methods of Analysis*, 18th edition. AOAC International, Gaithersburg, MD, USA, 79-80.

Cefas (2006). *The use of the Jellett Rapid Testing Kit for PSP Toxins Detection in the UK Statutory Monitoring Programme for Marine Toxins.* Cefas contract report.

Costa, P. R., Baugh, K. A., Wright, B., RaLonde, R., Nance, S. L., Tatarenkova, N., Etheridge, S. M. and Lefebvre, K. A. (2009). Comparative determination of paralytic shellfish toxins (PSTs) using five different toxin detection methods in shellfish species collected in the Aleutian Islands, Alaska. *Toxicon*, *54*, 313-320.

Dorantes-Aranda, J. J., Campbell, K., Bradbury, A., Elliott, C. T., Harwood, D. T., Murray, S. A., Ugalde, S. C., Wilson, K., Burgoyne, M. and Hallegraeff, G. M. (2017). Comparative performance of four immunological test kits for the detection of Paralytic Shellfish Toxins in Tasmanian shellfish. *Toxicon*, *125*, 110-119.

Dorantes-Aranda, J. J., Hayashi, A., Turnbull, A. R., Jolley, J. Y. C., Harwood, D. T. and Hallegraeff, G. M. (2020). Detection of paralytic shellfish toxins in southern rock lobster *Jasus edwardsii* using the qualitative Neogen™ lateral flow immunoassay: single-laboratory validation. *J. AOAC Int.*, *103*, 784–791.

Dorner, B. G., Zeleny, R., Harju, K., Hennekinne, J. A., Vanninen, P., Schimmel, H. and Rummel, A. (2016). Biological toxins of potential bioterrorism risk: Current status of detection and identification technology. *Trends Analyt. Chem.*, *85*, 889-102.

Eangoor, P., Indapurkar, A. S., Vakkalank, M., Yeh, J. S. and Knaack, J. S. (2017). Rapid and Sensitive ELISA Screening Assay for Several Paralytic Shellfish Toxins in Human Urine. *J Anal. Toxicol.*, *41*, 755-759.

EFSA (2009). European food safety authority. *EFSA J.*, *1019*, 1-76.

DeGrasse, S., Rivera, V., Roach, J., White, K., Callahan, J., Couture, D., Simone, K., Peredy, T. and Poli, M. (2014). Paralytic shellfish toxins in clinical matrices: extension of AOAC official method 2005.06 to human urine and serum and application to a 2007 case study in Maine. *Deep Sea Res. Part II Top. Stud. Oceanogr*, *103*, 368-375.

Gad, S. E. (2014). Saxitoxin. In: Wexler, P. (Ed.), *Encyclopedia of Toxicology*, 3rd edition. Academic Press, Oxford, 218-220.

García, C., Barriga, A., Díaz, J. C., Lagos, M. and Lagos, N. (2010). Route of metabolization and detoxication of paralytic shellfish toxins in humans. *Toxicon*, *55*, 135-144.

Harrison, K., Johnson, S. and Turner, A. D. (2016). Application of rapid test kits for the determination of paralytic shellfish poisoning (PSP) toxins in bivalve molluscs from Great Britain. *Toxicon*, *119*, 352-361.

ISO 16140:2003(E) (2003). *Microbiology of food and animal feeding stuffs-Protocol for the validation of alternative methods.*

ISO 16140-2:2016(E) (2016). *Microbiology of food chain-method validation-Part 2: Protocol for the validation of alternative (proprietary) methods against a reference method.*

Jellett, J. F., Roberts, R. L., Laycock, M. V., Quilliam, M. A. and Barrett, R. E. (2002). Detection of paralytic shellfish poisoning (PSP) toxins in shellfish tissue using MIST Alert™, a new rapid test, in parallel with the regulatory AOAC® mouse bioassay. *Toxicon*, *40*, 1407-1425.

Kawatsu, K., Kanki, M., Harada, T. and Kumeda, Y. (2014). A highly rapid and simple competitive enzyme-linked immunosorbent assay for monitoring paralytic shellfish poisoning toxins in shellfish. *Food Chem.*, *162*, 94-98.

Komano, A., Maruko, H. and Sekiguchi, H. (2011). Detection of saxitoxin in counterterrorism using a commercial lateral flow immunoassay kit. *Forensic Toxicol.*, *29*, 38-43.

Laycock, M. V., Donovan, M. A. & Easy, D. J. (2010). Sensitivity of lateral flow tests to mixtures of saxitoxins and applications to shellfish and phytoplankton monitoring. *Toxicon*, *55*, 597-605.

Laycock, M. V., Kralovec, J. and Richards, R. (1995). Some *in vitro* chemical interconversions of paralytic shellfish poisoning (PSP) toxins useful in the preparation of analytical standards. *J. Mar. Biotechnol.*, *3*, 121-125.

LeDoux, M. and Hall, S. (2000). Proficiency testing of eight French Laboratories in using the AOAC mouse bioassay for paralytic shellfish poisoning: Interlaboratory collaborative study. *J. AOAC Int.*, *83*, 305-310.

Lehane, L. (2000). *Paralytic shellfish poisoning: A review.* Canberra: National Office of Animal and Plant Health, Agriculture, Fisheries and Forestry – Australia.

Luckas, B., Hummert, C. and Oshima, Y. (2003). Analytical methods for paralytic shellfish poisons. In: Hallegraeff, G. M., Anderson, D. M., Cembella, A. D. (Eds.), *Manual on Harmful Marine Microalgae.*, UNESCO, France, pp. 191–209.

Mackintosh, F. H., Gallacher, S., Shanks, A. M. and Smith, E. A. (2002). Assessment of MIST Alert^{TM}, a commercial qualitative assay for detection of paralytic shellfish poisoning toxins in bivalve mollusks. *J. AOAC Int.*, *85*, 632-641.

McFarren, E. F. (1959). Report on collaborative studies of the bioassay for paralytic shellfish poison. *J. Assoc. Off. Anal. Chem.*, *42*, 263-271.

Oshima, Y. (1995). Post-column derivatization HPLC methods for paralytic shellfish poisons. In Hallegraeff, G. M., Anderson, D. M., Cembella, A. D. (Eds.), *Manual on Harmful Marine Microalgae.*, UNESCO, France, 81–94.

Park, D. L., Adams, W. N., Graham, S. L. and Jackson, R. C. (1986). Variability of mouse bioassay for determination of paralytic shellfish poisoning toxins. *J. Assoc. Off. Anal. Chem.*, *69*, 547–550.

Park, D. L., Guzman-Perez, S. E. and Lopez-Garcia, R. (1999). Aquatic biotoxins: design and implementation of seafood safety monitoring programs. *Rev. Environ. Contam. Toxicol.*, *161*, 157–200.

Pitschmann, V. and Hon, Z. (2016). Military Importance of Natural Toxins and Their Analogs. *Molecules*, *21*, 556.

Posthuma-Trumpie, G. A., Korf, J. and van Amerongen, A. (2009). Lateral flow (immuno)assay: its strengths, weaknesses, opportunities and threats. A literature survey. *Anal. Bioanal. Chem.*, *393*, 569–582.

Rapala, R., Robertson, A., Negri, A. P., Berg, K. A., Tuomi, P., Lyra, C., Erkomaa, K., Lahti, K., Hoppu, K. and Lepistö, L. (2005). First report of saxitoxin in Finnish lakes and possible associated effects on human health. *Envion. Toxicol.*, *20*, 331-340.

Reverté, L., Soliño, L., Carnicer, O., Diogène, J. and Campàs, M. (2014). Alternative methods for the detection of emerging marine toxins: biosensors, biochemical assays and cell-based assays. *Mar. Drugs*, *12*, 5719-5763.

Sajid, M., Kawde, A. N. and Daud, M. (2015). Designs, formats and applications of lateral flow assay: A literature review. *J. Saudi Chem. Soc.*, *19*, 689-705.

Turner, A. D., Stubbs, B., Coates, L., Dhanji-Rapkova, M., Hatfield, R. G., Lewis, A. M., Rowland-Pilgrim, S., O'Neil, A., Stubbs, P., Ross, S., Baker, C. and Algoet, M. (2014). Variability of paralytic shellfish toxin occurrence and profiles in bivalve molluscs from Great Britain from official control monitoring as determined by pre-column oxidation liquid chromatography and implications for applying immunochemical tests. *Harmful Algae*, *31*, 87-99.

Turner, A. D., Tarnovius, S., Johnson S., Higman W. A. and Algoet, M. (2015). Testing and application of a refined rapid detection method for paralytic shellfish poisoning toxins in UK shellfish. *Toxicon*, *100*, 32-41.

Turner, A. D., Tarnovius, S., Hatfield, R. G., Teixeira Alves, M., Broadwater, M, Van Dolah, F., Garcia-Mendoza, E., Medina, D., Salhi, M., Goya, A. B., Barrera, F., Carrasco, D., Rubilar, I. and Suarez-Isla, B. A. (2020). Application of Six Detection Methods for Analysis of Paralytic Shellfish Toxins in Shellfish from Four Regions within Latin America. *Mar Drugs*, *18*, 616.

Usleber, E., Dietrich, R., Burk, C., Schneider, E. and Märtlebauer, E. (2001). Immunoassay methods for paralytic shellfish poisoning toxins. *J. AOAC Int.*, *84*, 1649–1656.

Vale, C., Alfonso, A., Vieytes, M. R., Romaris, X. M., Arevalo, F., Botana, A. M. and Botana, L. M. (2008). *In vitro* and *in vivo* evaluation of paralytic shellfish poisoning toxin potency and the influence of the pH of extraction. *Anal. Chem.*, *80*, 1770-1776.

Wharton, R. E., Feyereisen, M. C., Gonzalez, A. L., Abbott, N. L., Hamelin, E. I. and Johnson, R. C. (2017). Quantification of saxitoxin in human blood by ELISA. *Toxicon*, *133*, 110-115.

Wiese, M., D'Agostino, P. M., Mihali, T. K., Moffitt, M. C. and Neilan, B. A. (2010). Neurotoxic alkaloids: saxitoxin and its analogs. *Mar. Drugs*, *8*, 2185-2211.

Wong, C. K., Hung, P., Ng, E. A. L., Lee, K. L. H., Wong, G. T. C. and Kam, K. M. (2010). Operational application of a rapid antibody-based detection assay for first line screening of paralytic shellfish toxins in shellfish. *Harmful Algae*, *9*, 636-646.

Wong, C. K., Hung, P., Ng, H. C. C., Lee, S. Y. and Kam, K. M. (2011). Cluster analysis of toxins profile pattern as a tool for tracing shellfish contaminated with PSP-toxins. *Environ. Res.*, *111*, 1083-1090.

In: ELISA
Editor: Jose Ma. M. Angeles
ISBN: 978-1-68507-500-2

Chapter 7

APTAMERS: THE CHEMICAL ANTIBODIES, THEIR EMERGENCE, AND APPLICATION TO ENZYME-LINKED IMMUNOSORBENT ASSAY

***Altair B. Agmata**, *Chembie A. Almazar**, *Marjo V. Mendoza** and *Windell L. Rivera*†**
Pathogen-Host-Environment Interactions Research Laboratory,
Institute of Biology, College of Science, University of the Philippines
Diliman, Quezon City, Philippines

ABSTRACT

The advancement of *in vitro* methods to select high-affinity ligands through combinatorial chemistry approaches has paved the way for the recent development of aptamers in biosensing technologies. Aptamers are single-stranded oligonucleotides with high specificity and affinity (at the nanomolar or picomolar scale) to target a wide variety of

* These authors contributed equally to this study.
† Corresponding Author's E-mail: wlrivera@science.upd.edu.ph.

molecules ranging from small organic molecules to whole cells. With substantial attention being paid to the development of aptamers, the reliance of biosensing approaches on antibodies as molecular recognition elements (MREs) is now changing. For example, enzyme-linked aptamer assays (ELAA) were recently developed from the well-established enzyme-linked immunosorbent assays (ELISA), exploiting aptamers' stability, flexibility, and reproducibility compared with antibodies. In this chapter, the basic characteristics of aptamers and their unique properties compared with antibodies are summarized. Moreover, ELAA's merit as an analytical biosensing platform is examined while exploring some of the modes in which the assay is commonly implemented, along with their corresponding advantages and disadvantages. Recent progress and applications of ELAA as well as the extension of aptamers to other biosensing platforms are also reviewed, highlighting aptamers' potential and impact on various fields, including medical diagnostics, therapeutics, food analysis and safety, and environmental monitoring.

Keywords: aptamers, diagnosis, ELAA

INTRODUCTION

Nucleic acids (DNA and RNA) have long been known as the genetic blueprint of life, whose major functions include the storage and expression of genetic information. However, their ability to fold into complex three-dimensional structures led to the discovery of their involvement in enzymatic activity, regulation of gene expression, cellular responses, and protein synthesis. These insights into the multifunctional nature of nucleic acids paved the way for aptamer research (Deigan and FerrÉ-D'AmarÉ 2011; Zemora and Waldsich 2010).

Investigations on HIV revealed that small RNA regions, known as trans-acting response elements (TAR), can recognize and bind to specific viral and cellular proteins (Sullenger et al. 1990, 1991). Thus, the concept of using nucleic acids as affinity or recognition molecules targeting specific proteins emerged. First defined in 1990, the term "aptamer," which comes from the Latin word *aptus* meaning "to fit," describes the ability of aptamers to change their structure in order to bind to their target

(Ellington and Szostak 1990; Tuerk and Gold 1990). Aptamers are high-affinity nucleic acid molecules comprising short, single-stranded DNA or RNA, usually 20–80 nucleotides in length. They can bind to virtually any type of target from ions and small organic or inorganic molecules to large protein complexes, whole cells, and even live animals by folding into unique tertiary structures through chemical modification (Zhou and Rossi 2017). The potential for using aptamers in various research fields has been extensively studied, particularly in clinical diagnostics and therapeutics, in which they are regarded as possible alternative to antibodies. Relative to antibodies, aptamers offer several advantages, such as their small size, better stability, higher affinity and specificity to different chemical and biological targets, modifiability, higher reproducibility with minimal batch-to-batch variation, and low production cost (Mutreja et al. 2019; Zhang et al. 2019). All of these characteristics make aptamers ideally suited to replace antibodies as molecular probes in diagnostics, biomedical imaging, therapeutics, and biosensing.

APTAMER GENERATION

Aptamers are commonly generated through a process called the systematic evolution of ligands by exponential enrichment (SELEX) (Ellington and Szostak 1990; Tuerk and Gold 1990). This selection approach generally consists of three critical steps: (i) incubation, (ii) partitioning, and (iii) recovery and re-amplification (Figure 1). The cycle starts with incubation of the target with an oligonucleotide library composed of random single-stranded nucleotides (20–80-mer) flanked by primer binding sites at the 5' and 3' ends. Bound oligonucleotides are then partitioned or separated from unbound components of the library. Finally, target-bound oligonucleotides are eluted and re-amplified by PCR to create a new library for the next cycle. This process is repeated (8–15 cycles) until aptamers with the desired affinity and specificity are enriched and sequenced (Kulabhusan et al. 2020; Song et al. 2012). The major difference between DNA and RNA SELEX methods is that RNA SELEX requires *in vitro* transcription. RNA aptamers are initially amplified by T7 RNA polymerase and reverse-transcribed into double-stranded DNA prior to PCR (Vorobyeva et al. 2018; Urak et al. 2016).

Over the years, variations of this methodology have been developed to generate functional candidate aptamers with strong affinity and improved selectivity. Among these approaches are counter SELEX, negative SELEX, immunoprecipitation-coupled SELEX (IP-SELEX), capture SELEX, whole-cell SELEX, capillary electrophoresis SELEX (CE-SELEX), atomic force microscopy SELEX (AFM-SELEX), and live-animal SELEX (Zhang et al. 2019; Wu, J. et al. 2014).

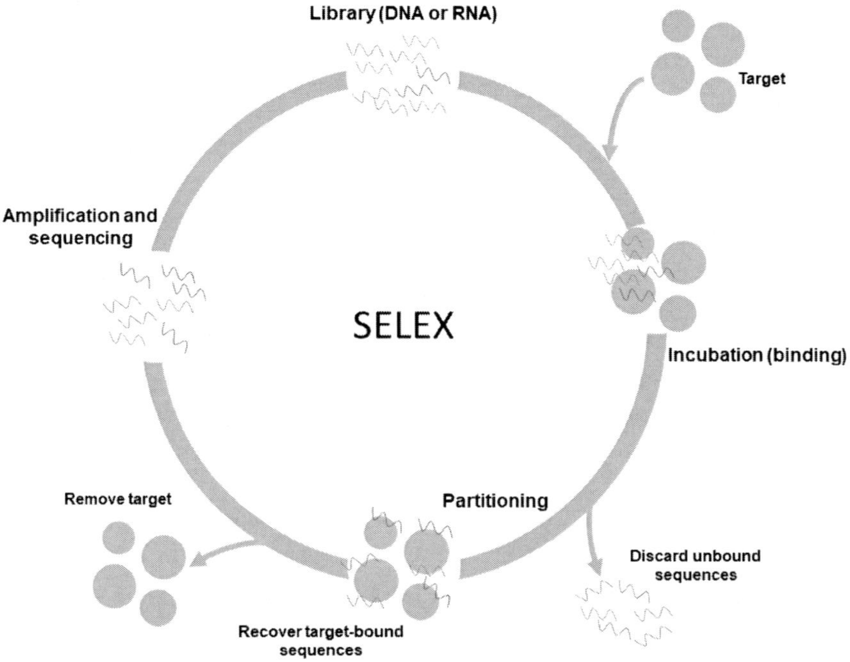

Figure 1. SELEX procedure.

APTAMER VS. ANTIBODY

Aptamers are also known as chemically synthesized antibodies since they mimic antibodies' ability to bind to a wide range of targets. In recent years, the amount of aptamer research has rapidly increased as promising results emerged about their potential ability to replace traditional antibodies, specifically in immunoassays and therapeutics (Kulabhusan et al. 2020; Nezlin 2016). Major benefits of using aptamers over antibodies are described in Table 1. The key advantages of aptamers include – that they are inherently small (approximately 5–15 kDa) and far less immunogenic, facilitating faster infiltration and better access to targets unreachable by antibodies. Additionally, this makes their synthesis and chemical modification easier. However, owing to their size, aptamers are likely to be excreted from the bloodstream via renal filtration, which poses a major challenge in therapeutics (Lakhin et al. 2013). This matter was addressed by conjugation experiments wherein extended bloodstream circulation of aptamers was observed post-modification at certain positions, for example, base, 2'- sugar ring, and 3'-end. Aptamers conjugated with nanoparticles, polyethylene glycol (PEG), and cholesterol have shown improved tissue penetration, and decreased renal filtration and blood clearance (Yu et al. 2010; Ng et al. 2006; Rusconi et al. 2004).

Antibodies are known to be easily denatured and are sensitive to high temperatures, thus requiring refrigeration for storage and transport. Their denaturation, which may result from a change in pH, temperature, or salt concentration, is irreversible, and they are also susceptible to proteases, subsequently reducing their shelf-life. Although antibodies have been widely used, these limitations remain key issues in their biomedical application (Ma et al. 2020). Aptamers are more robust than antibodies, exhibiting better thermal and conformational stability. Heat denaturation of aptamers may also occur, but despite this, their denaturation is reversible, and aptamers can recover their native conformation without loss of their activity (Song et al. 2012). Nuclease degradation has also been a major issue in the production of aptamers. Several techniques have been introduced to resist the degradation of aptamers by nucleases, such as synthetically combining moieties by sugar modifications, biotin conjugation, and inversion of the dT cap to increase stability while maintaining specificity and affinity. These approaches have been successful in producing nuclease-resistant functional aptamers (Kratschmer and Levy 2017; Kuwahara and Sugimoto 2010; Healy et al. 2004). Such features contribute to the longer shelf-life and stability of aptamers, which are important in most immunoassays and other related applications requiring a wide range of assay conditions.

The successful development of aptamers lies in their surface recognition ability. Aptamers and antibodies bind to specific epitopes or regions of target antigens. With specific epitope binding, aptamers and antibodies can effectively detect their target. Cross-species affinity occurs when identical epitopes are conserved among different types of antigens, for example, among species of pathogen. As there is a wide array of existing antigens, antibodies need to be versatile enough to recognize their targets. Some aptamers have been designed to identify single point mutations and conformational isomers. Advancement in aptamer engineering has led to the successful development of an RNA aptamer that can identify a point mutation in codon 12 of KRAS, a gene associated with cancer (Jeong et al. 2010). On the other hand, an aptamer capable of selectively binding to a single isomeric form has been developed by incorporating photoisomerization in the *in vitro* selection process of an RNA aptamer (Young and Deiters 2008). These new aptamer variants with enhanced selectivity have a vast range of potential applications not only in diagnostics but also in other biomedical fields.

In principle, both aptamers and antibodies have high affinity and selectivity. The high affinity of aptamers or antibodies for their target is a result of binding through non-covalent interactions, which are mainly via van der Waals forces, as well as hydrogen bonding and electrostatic interactions. Antibody or aptamer binding is greatly dependent on the number of identical epitopes on the antigen (Zhou and Rossi 2017; Sela-Culang et al. 2013). However, several studies have shown that aptamers can be designed to have better affinity and selectivity than antibodies due to their conformational adaptability and flexibility. To enhance aptamer binding affinity, several approaches have been explored, including the stabilization of aptamer conformations, sequence optimization, and construction of multivalent aptamers (Adachi and Nakamura 2019; Hasegawa et al. 2016). Specifically, truncation of the aptamer stem-loop regions yielded substantially higher binding affinity, as performed in a cancer cell detection assay (Kaur and Yung 2012). Combinatorial screening methods using microarrays, blotting techniques, and genetic algorithms were performed to produce high-affinity aptamers (Yoshida et al. 2009; Asai et al. 2004). In fact, the use of an array-based platform has revealed high-affinity aptamer pairs that bind to distinct sites on their protein target, human angiopoietin-2. This technique was adopted from antibody pairing used to increase the specificity of some immunoassays (Cho et al. 2015). The construction of multivalent aptamers through the concatenation of identical or different aptamer motifs allowed binding to multiple target sites, thus increasing their affinity (Vorobyeva et al. 2018).

Antibodies have played an integral role in most diagnostic procedures, particularly in immunoassays, and in therapeutic research. However, the generation of antibodies is performed *in vivo*, which is often laborious, costly, and time-consuming. Their targets must also have a strong immune response, thus limiting the number of possible targets (Zhang et al. 2019; Nezlin 2016). The application of cell culture systems has advanced the large-scale production of antibodies; however, batch-to-batch consistency and purification remain to be the ongoing challenges in antibody production (Shukla et al. 2017). All of these factors are driving aptamer research to overcome such limitations of antibodies. Aptamers are chemically synthesized *in vitro* through combinatorial chemistry methodologies, the most common of which is via SELEX; hence, there is less risk of contamination and lower batch-to-batch variation. The production of aptamers is also cost-efficient and can be done within weeks (Zhang et al. 2019; Zhou and Rossi 2017). Perhaps the major advantage of aptamers over antibodies is that they can be selected to bind to a wide range of targets including non-immunogenic small molecules that antibodies cannot recognize, as well as toxins, pathogens, nucleic acids, proteins, whole cells, tissues, and live animals (Kulabhusan et al. 2020; Sola et al. 2020).

Table 1. Comparison between aptamers and antibodies

	Aptamer	Antibody
Size	Small molecules (~5–15 kDa), better access to their target	Relatively big (~150–180 kDa), resist renal filtration
Stability	Thermally stable Denaturation is reversible Degradable by nucleases Longer shelf-life	Temperature-sensitive requiring refrigeration Denaturation is irreversible Degradable by proteases Limited shelf-life
Affinity and specificity	High target affinity and specificity Multivalent aptamers increase affinity to target Identifies single point mutations and conformational isomers	High affinity and specificity but dependent on the number of identical epitopes on the antigen
Immunogenicity	Less immunogenic and non-toxic	Immunogenic unless humanized
Target	Multiple, from ions and small molecules to cells and live animals	Limited to immunogenic proteins or molecules, mainly including cells, viruses, and nucleic acids
Synthesis and modifications	*In vitro* preparation through chemical synthesis Aptamers can be selected under non-physiological conditions Chemical modifications can be done easily without loss of function	*In vivo* production; cell culture system Modifications may affect binding affinity
Batch variation	High reproducibility with low batch-to-batch variability	High batch-to-batch variation
Production cost	Low	High

Enzyme-Linked Aptamer Assay (ELAA)

The enzyme-linked aptamer assay (ELAA) is based on the well-established enzyme-linked immunosorbent assay (ELISA), where antibodies are replaced with aptamers as the molecular recognition element (MRE). ELAA is also known for its variants, including enzyme-linked apta-sorbent assay (ELASA), enzyme-linked oligonucleotide assay (ELONA), and aptamer-linked immobilized sorbent assay (ALISA). It exploits aptamers' superiority over antibodies in terms of stability, flexibility, reusability, and reproducibility. The ability to use aptamers in analytical applications such as ELAA was driven by the advent of combinatorial chemistry methodologies (Gordon et al. 1994); enabling the rapid selection of high-affinity ligands including nucleic acids through the SELEX approach. This was first demonstrated by Drolet et al. (1996) through the use of ELONA to quantitatively detect human vascular endothelial growth factor (hVEGF). The results of their study demonstrated that ELONA's accuracy, precision, and specificity are comparable to the limits achieved by conventional sandwich ELISA assays, establishing the viability of aptamers as an alternative MRE. Some of the key features of ELISA and ELAA as analytical assays are compared in Table 2.

Table 2. Summary of key features of ELAA vs. ELISA as assays (Toh et al. 2015)

	ELAA	ELISA
Sensitivity	Similar	Similar
Automation	Yes	Yes
Temperature requirements	None; can withstand up to ~60°C	4°C is required for antibodies
Shelf-life	Longer	Short
Cost	Lower	Higher
Reusability	Yes	No

Similar to ELISA, ELAA is commonly classified into three main categories, namely, direct, indirect, and sandwich configurations (Figure 2). Both the indirect and sandwich configurations allow the complementary use of antibodies, enabling more flexibility. The configuration of ELAA to be used mainly depends on the availability of other complementary binding agents besides aptamers that allow attachments to various regions of the target's surface (Toh et al. 2015).

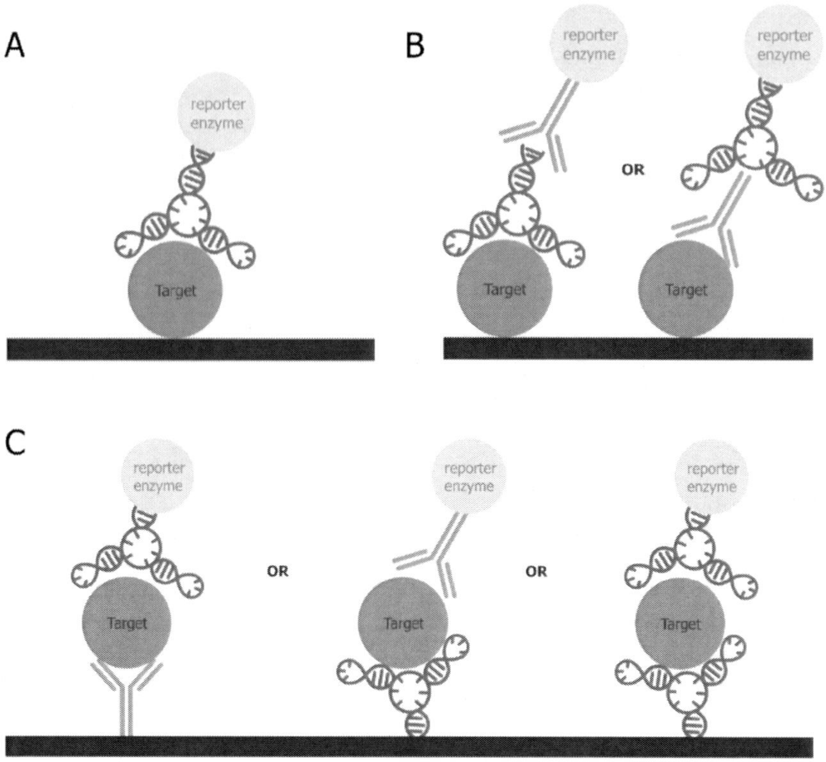

Figure 2. Various ELAA configurations: A) direct ELAA, B) indirect ELAA, and C) sandwich ELAA (including antibody-target-aptamer, aptamer-target-antibody, and aptamer-target-aptamer modes).

Direct ELAA

Direct ELAA is considered the simplest and fastest method among all formats of ELAA (Toh et al. 2015). In this configuration, the target analyte is immobilized on the surface of the platform. Biotinylated aptamers are then added to allow binding with the target analyte, followed by the washing out of unbound aptamers. The quantity of bound aptamer is then detected by conjugation with a reporter protein, typically horseradish peroxidase, via its streptavidin moiety. In many cases, a blocking step is performed, aimed at ensuring saturation of the immobilization surface. This leads to improved sensitivity and specificity by avoiding background interference and non-specific interactions. Numerous studies (Ghavamipour et al. 2020; Wildner et al. 2019; Lamont et al. 2011; Balogh et al. 2010; Vivekananda and Kiel 2006) have used bovine serum albumin (BSA) in blocking solutions, due to its effectiveness in biotin-streptavidin systems (Vargas-Montes et al. 2019).

Owing to its simplicity and speed, direct ELAA is the usual format of choice for estimating the dissociation constants (*Kd*) of aptamers (Toh et al. 2015). *Kd* is a measure of aptamers' binding affinity to their target at chemical equilibrium. It physically corresponds to the concentration of aptamer in solution capable of saturating half of the total target. This interpretation is further demonstrated in the equations below:

$$K_d = \frac{[A][T]}{[AT]} \tag{1}$$

where [A], [T], and [AT] are aptamer, target, and bound aptamer–target concentrations, respectively. At half target saturation$\left(\frac{[AT]+[T]}{[T]} = \frac{1}{2}\right)$, $[T] = [AT]$, thus:

$$K_d = \frac{[A][AT]}{[AT]} = [A] \tag{2}$$

The interpretation above suggests that the lower the *Kd* value, the higher the affinity of the aptamer to its target; hence, a low *Kd* value is desirable. *Kd* is experimentally determined through non-linear curve-fitting of absorbance values obtained from ELAA experiments (Figure 3). The model typically used in various studies (Vargas-Montes et al. 2019; Song et al. 2017; Wongphatcharachai et al. 2013) is the Hill equation (Hill 1910) given below:

$$y = y_{max} \frac{[A]}{K_d+[A]} \tag{3}$$

where y is a measure of binding concentration, typically absorbance or optical density, and y_{max} is its maximum value.

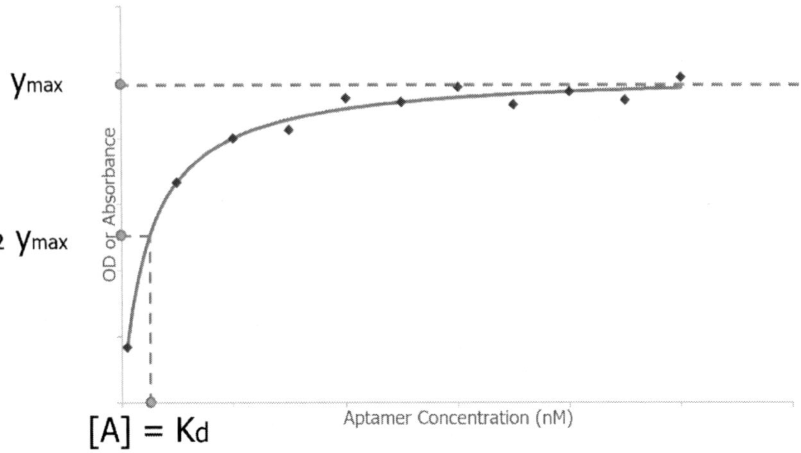

Figure 3. A typical non-linear regression curve for the determination of *Kd* through ELAA binding experiments.

INDIRECT ELAA

Indirect ELAA is similar to direct ELAA in that both modes immobilize the target on the platform surface. The difference lies in the usage of a secondary MRE as an agent for detecting where the reporter protein is attached. The configuration permits the usage of either an aptamer or an antibody as both primary and secondary MRE, allowing more flexibility depending on the availability of binding agents. Indirect ELAA is not very popular compared with the sandwich mode since target immobilization is almost limited to physical adsorption. This type of surface immobilization can be non-specific and may allow the recruitment of impurities along with the target analyte on the platform surface, especially at low target concentrations where impurities may outcompete the target (Toh et al. 2015). However, the format can be more sensitive than direct ELAA since it is ensured that the reporter protein moiety will not interfere with the primary MRE's conformation and thus its binding affinity (Baldrich et al. 2005b), especially for aptamers for which folding and tertiary structure, are key to strong and specific target binding.

SANDWICH ELAA

Among the other modes, the sandwich format is the most commonly used type of ELAA (Jarczewska and Malinowska 2020). The configuration is called "sandwich" as the target analyte is bound between either an aptamer or an antibody, both of which bind on different surface regions on the target. Sandwich ELAA is known to perform better than direct and indirect modes as it depends on the effectiveness of the capturing agent rather than the method of target immobilization, purity, or concentration (Marnissi et al. 2020; Toh et al. 2015). The sandwich mode also allows the use of aptamer-antibody hybrid assays, which reportedly provide lower limits of detection (Jarczewska and Malinowska 2020; Baldrich et al. 2005b).

The sandwich mode requires immobilization of the capturing agent (antibody or aptamer) onto the platform surface and is a critical aspect of the assay's performance. Antibodies are more popular capturing agents than aptamers since the latter are more easily labeled and labeling also has no significant effect on the aptamer's conformation (Toh et al. 2015). In such cases, antibody surface immobilization techniques used in a typical ELISA sandwich assay are implemented as both ELAA and ELISA assays are similar in this respect. These typically rely on simple surface physical adsorption through hydrophobic and electrostatic interactions, and hydrogen bonding on appropriate solid surfaces. Although this approach provides the simplest route, antibodies on the surface can be denatured, aggregated, displaced in subsequent washing steps, and immobilized in random orientations (Welch et al. 2017; Butler 2000). Meanwhile, covalent immobilization techniques offer improved performance as compared to simple physical adsorption but require further platform surface engineering to facilitate covalent bonding between the antibody and the surface. Surfaces where antibodies are commonly immobilized include gold, quartz, silica, polystyrene, polyacrylamide, Sepharose, agarose, cellulose, dextran, hydroxyapatite, magnetite, steel, and niobium oxide (Jarczewska and Malinowska 2020).

Similarly, aptamer immobilization techniques are also based on physical adsorption and covalent tethering. Aptamer physical adsorption commonly exploits the aptamer's negatively charged phosphate backbone to facilitate electrostatic interaction with cationic surfaces such as polyethylenimine, cysteamine, and chitosan (Jarczewska and Malinowska 2020). However, aptamer immobilization on cationic surfaces may interfere with proper aptamer folding, causing loss of its ability to molecularly recognize the intended target (Urmann et al. 2017). Hence, covalent linkage of aptamers on various surfaces is the more favored method. Aptamer covalent immobilization can be implemented through the direct attachment of aptamers with appropriate linkers, such as thiol-terminated ones (Balamurugan et al. 2008b).

Immobilization on functionalized surfaces is also applied through biotinylated aptamers that can be conjugated to surface-immobilized streptavidin (Jarczewska and Malinowska 2020). Additionally, spacers are sometimes added to the immobilized aptamers to decrease steric strain and facilitate proper aptamer folding. An important feature is the aptamer's reliance on a proper tertiary structure, rather than the sequence, as in the case of DNA probes, to facilitate effective recognition and binding to its target (Baldrich et al. 2005b). Indeed, several studies (Balamurugan et al. 2008a; Centi et al. 2007; Baldrich et al. 2005a; Bladrich et al. 2005b) have demonstrated the significance of spacers in increasing the binding affinity of aptamers to their targets. Various molecular chains are used as aptamer spacers, including polynucleotide (poly-T and poly-A chains) and carbon chains.

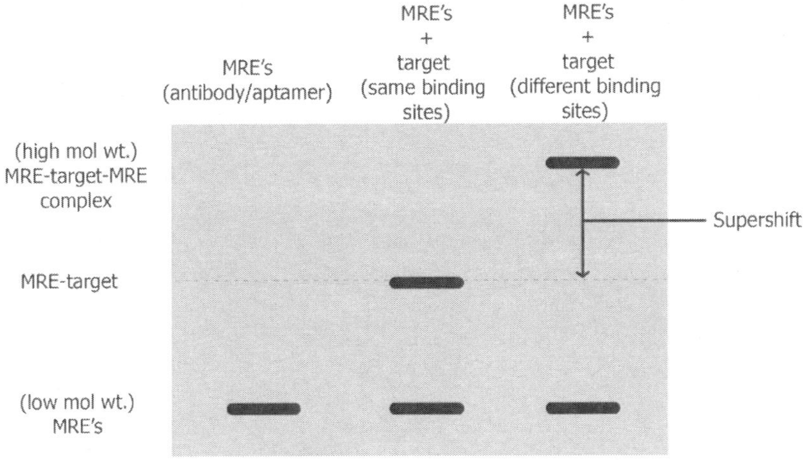

Figure 4. EMSA and the presence of a supershift for the confirmation of two different binding sites for two distinct MREs (antibody/aptamer).

Notably, the sandwich assay requires two target binding regions, one for capturing and the other for the detecting agent. It is important that these two binding regions do not overlap; in other words, the two MREs to be used should not bind at the same site. To ensure that this requirement is met, a gel electrophoretic mobility shift assay (EMSA) can be performed to look for the presence of a supershift (Toh et al. 2015). The presence of a supershift, a band corresponding to the heaviest complex with the slowest gel migration, indicates an MRE-target-MRE complex. Such a complex with a greater molecular weight than the bimolecular MRE-target complex is only possible if the two distinct MREs are capable of binding to two distinct target regions (Figure 4). Moreover, with the advent of *in silico* molecular modeling approaches, the relative thermodynamic stability of each MRE to a specific target site can also be explored through molecular dynamics or molecular docking (Figure 5). Such simulations can aid in making inferences about whether candidate MREs would compete with each other on the same binding site, or if each MRE will have distinct stable binding sites.

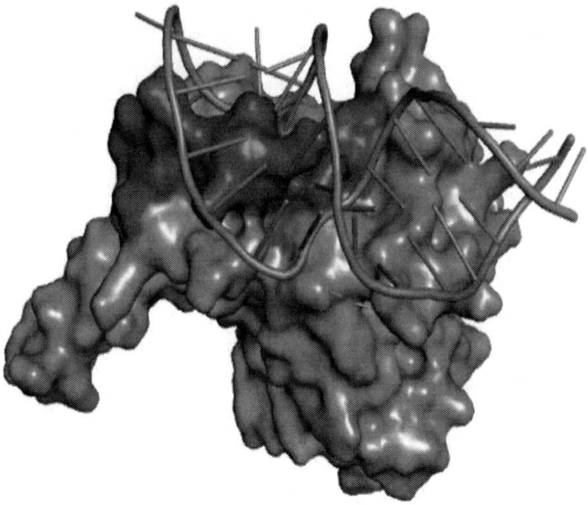

Figure 5. Cartoon and surface representation of a 3D model of a hypothetical aptamer docked to an epitope target site on LipL32A protein simulated by the authors through HADDOCK (van Dijk et al. 2006).

APPLICATION OF APTAMERS

Aptamers form three-dimensional structures that can attach to a variety of specific targets with high affinity and specificity. Compared with antibodies, aptamers can be synthesized chemically, eliminating the use of cell lines and decreasing the cost of production. In addition, they exhibit extremely low variability between batches and are easier to modify post-production. They are also more stable and can return to their original conformation following denaturation without loss of activity and are minimally immunogenic. Aptamers can be applied in a variety of fields, including diagnostics, therapeutics, and environmental monitoring such as food inspection and hazard detection.

APTAMERS IN DIAGNOSTICS

Aptamers in diagnostics provide robust methods to detect biomarkers that have a relatively low abundance and thus have the potential to be applied to point-of-care testing systems. One study by Nazari et al. (2019) developed an aptamer-based sandwich assay for measuring thymidine kinase (TK1) in cancer patients. TK1 is an excellent biomarker that is elevated during the development of malignancies. ELAA has enabled the detection of TK1 in the concentration range of 54–3500 pg mL^{-1}. Another point-of-care application of ELAA is in the detection of neutrophil gelatinase-associated lipocalin (NGAL) in patients with acute kidney disease. The candidate aptamers can detect the biomarker at a minimum concentration of 30.45 ng/mL, allowing early diagnosis of the disease, which also have further applications as currently being tested in clinical trials (Hong et al. 2019). Aptamers have also been useful for detecting certain molecular biomarkers for diagnostics concerning the central nervous system. The detection of dopamine in the blood has been established by an RNA-based ELAA that is especially helpful to patients with ADHD and Parkinson's (Park and Paeng 2011). Aside from dopamine, much research on aptamer technology has also extended to the detection of different types of neurotransmitters that affect mood and behavioral disorders, which have potential for early diagnosis and possible treatment (Sinha and Mukhopadhyay 2020).

Emerging infectious diseases have become an increasing problem, especially in developing countries. The surveillance of pathogens is of utmost importance for the recognition, control, and management of these diseases. Aptamers offer a promising approach to detect pathogens in the environment, food, and clinical settings. This technique can be quite practical due to its sensitivity, as well as rapid and efficient detection that eliminate the need for laborious conventional methods of bacterial culture and bacterial identification.

The most widely applied aptamer-based methods for detecting bacteria employ nanoparticle-based colorimetric assays, where aptamers bind and capture specific organisms and the accumulation of nanoparticles is associated with a change in color (Yuan et al. 2014). Several iterations of this technology include modification of the aptamers or nanoparticles to improve detection and sensitivity. One such modification includes coupling the aptamer nanoparticle with positive di-electrophoresis (pDEP) for the direct identification of microorganisms using fluorescence microscope imaging (Shangguan et al. 2015). These aptamers conjugated to fluorescent nanoparticles have also been used for flow cytometry assays on organisms with high sensitivity to reduce false positives (Borsa et al. 2016).

Enzyme-linked aptamer assays have also been established for the rapid detection of bacterial pathogens including *Salmonella* spp. (Ma et al. 2014), *Staphylococcus aureus* (Yuan et al. 2014), *Streptococcus* spp. (Alfavian et al. 2017), and *Escherichia coli* (Marton et al. 2016) using the cell SELEX procedure. Moreover, aptamers have been developed to increase affinity and specificity such that molecular probes are generated to detect viral pathogens like the vaccinia virus (Parekh et al. 2010), hepatitis B (Suh et al. 2014) and C viruses (Pleshakova et al. 2018), human immunodeficiency virus (Tombelli et al. 2005), Zika virus (Lee and Zeng 2017), and influenza virus (Wang et al. 2013).

Aside from bacterial and viral pathogens, aptamers have also been designed to detect several parasites such as *Entamoeba histolytica*, *Plasmodium* spp., and *Trypanosoma* spp. by binding to specific surface proteins (Ospina-Villa et al. 2016). A study by Vargas-Montes et al. (2019) utilized ELAA to detect *Toxoplasma gondii* in human blood serum. The assay detected the virulence protein ROP18 at a minimum concentration of 1.56 μg/mL, indicating significant value for the detection of severe congenital toxoplasmosis. Another study by Espiritu et al. (2018) utilized ELAA to detect *Trichomonas vaginalis*, a sexually transmitted pathogen that is difficult to diagnose in clinical settings. The designed aptamers could detect its adhesion protein, AP65, with no cross-reactivity to other urogenital microorganisms, demonstrating the high specificity of aptamers in rapid detection assays.

Most of these sandwich assays use DNA/RNA aptamers that have been immobilized in a streptavidin-coated matrix and conjugated with different materials, which allows detection of the capture of the target molecule. Other later studies employed different techniques to capture, bind, and express the presence of the analyte, where aptamers behave as biosensors called aptasensors. These aptasensors are then classified into different categories based on the optical principles used for the detection of the molecule (Zou et al. 2019).

APTAMERS IN THERAPEUTICS

Aptamers are promising drug candidates and are increasingly being studied in all phases of drug discovery and development. These small molecules can replace conventional antibodies as diagnostics and therapeutic tools in treating and managing diseases. Since aptamers can be chemically modified, for example, via the attachment of functional groups, chemicals, or nanoparticles, they make the perfect candidates for diagnostic applications (Toh et al. 2015). Aptamers currently in various phases of clinical trials include treatments for retinopathies, cancer, heart diseases, diabetes, various blood disorders, and renal diseases (Kulabhusan et al. 2020).

Pegaptanib (Macugen) is the first therapeutic aptamer to be approved by the US FDA (in 2004), which is an angiogenic medicine for treating age-related macular degeneration by intravitreal injection. Pegaptanib is a pegylated 27-nucleotide aptamer that specifically binds to $VEGF_{165}$, which blocks its interaction with its receptor, thereby preventing blood vessel formation during tumorigenesis (Ferrara 2004). This aptamer has been chemically modified by attaching 2'-fluoropyrimidines, 2'-O-methyl purines, a 3′-linked deoxy thymidine residue, and coupled to a 40 kDa polyethylene glycol (PEG) to reduce nuclease degradation and improve its half-life *in vivo*. This drug is currently approved for therapeutic use in the United States, Europe, Canada, Brazil, and Australia (Vinores 2006).

Since conventional therapeutic drugs for cancer are not selective and have caused serious toxic side effects, aptamers targeting cancer biomarkers that target specific tumor cells have been developed and screened. These aptamers can also act as drug delivery systems by coupling drugs, siRNA, and nanoparticles, among others, to form a complex that allows the targeted drug delivery system to reach cancer cells, reducing the toxic side effects, overcoming drug resistance, and increasing therapeutic efficacy (Kaur 2018).

The aptamers AS1411 (developed by Antisoma) and olaptesed pegol (also known as NOX-A12, produced by NOXXON Pharm) are right now in the stage II clinical phase (Rosenberg et al. 2014). AS1411, an unmodified DNA aptamer with a G-quadruplex design known as a non-SELEX aptamer, binds to nucleolin and has been shown to inhibit the growth of a wide range of malignant cells *in vitro* (Bates et al. 2009). Cell proliferation could likewise be inhibited by AS1411 by interrupting nuclear factor-κB (NF-κB) and destabilizing BCL-2 mRNA (Soundararajan et al. 2008).

To increase the ability of aptamers to specifically target cancer cells, scientists have developed and improved upon this technology by conjugating aptamers with different molecules. AptDC (aptamer-drug conjugate) models have been established and undergone preclinical evaluation. Aptamers conjugated with chemotherapeutic drugs have also been developed to better target cancer cells while decreasing the cytotoxic effects on normal cells. One of these AptDC models is an aptamer conjugated to gold nanoparticles and anchored to a cell surface protein (PrP^C) for the targeted delivery of doxorubicin to colorectal cancer cells. The therapeutic effects of this type of treatment include the inhibition of mitochondrial functions, which compared to Dox treatment alone increased apoptosis and greatly reduced colorectal cancer cell proliferation (Go et al. 2021). Another aptamer bound to siRNA specifically targets HER2 (human epidermal growth factor) in breast cancer cells. The aptamer-siRNA complex specifically targets and silences the X-box-binding protein (XBP1) in HER2+ mouse models, which dramatically improves prognosis by inhibiting cell proliferation and impairing the process of angiogenesis (Zhang et al. 2020).

Despite the broad range of antibiotics available, the emergence of multidrug-resistant bacteria remains a threat to public health. Aptamers, in conjunction with other antibiotics, have been developed to counter the resistance developed by bacteria. In 2011, Schlesinger et al. selected DNA aptamers to counter the activity of β-lactamase developed by certain bacteria that have become resistant to β-lactams, a class of antibiotics. The combination of aptamers and a β-lactam suppressed bacterial growth compared with antibiotic treatment alone. Another strategy involves the selection of aptamers that target bacterial components to stimulate passive immunity. DNA aptamers were developed against lipopolysaccharides from pathogenic *Escherichia coli* O111:B4 coupled to a human C1qrs, a part of the complement system, triggering the destruction of bacteria with little to no resistance (Bruno et al. 2008).

MONITORING ENVIRONMENTAL CONTAMINATION AND HAZARD DETECTION

Aptamers have the ability to detect, monitor, and minimize environmental pollutants that can be toxic to humans and the local fauna and flora. These toxic compounds include antibiotics, heavy metals, and toxins that can cause severe deficits to the nervous, endocrine, and reproductive systems (Mahurpawar 2015). To address these issues, several studies have evaluated the use of aptamers against known antibiotics like tetracycline and chloramphenicol (Mehta et al. 2011). Tetracycline is a broad-spectrum antibiotic that eventually accumulates in food products because of its uncontrolled use in farms. Aptamer technology has been improved in this regard by the hybridization of DNA aptamers to probes that produce a signal that can be detected using fluorescence, among other approaches, assisted by a nanomaterial to increase sensitivity and efficiency (Berens et al. 2001). Another study by Gong et al. (2019) developed the colorimetric detection of tetracycline on an ELAA with a multivalent HRP-mimicking DNAzyme for substrate color rendering. This technique improved the sensitivity to tetracycline with a limit of detection of 10^{-2} ng/mL^{-1}, even in complex samples like milk.

Heavy metal detection, on the other hand, benefits from the use of aptamers by recognizing trace heavy metal ions from various environmental sources. These molecules can bind to the bases of the DNA aptamer, forming stable complexes (Farzin et al. 2017). To date, several DNA aptamers that can detect heavy metal ions like Pb^{2+} (Burgstaller et al. 2002), Hg^{2+} (Qi et al. 2020), Cd^{2+} (Wu, Y. et al. 2014), and As^{3+} (Mao et al. 2020) have been developed. In the same vein, aptamers to detect toxic chemicals from herbicides and insecticides like atrazine (Abraham et al. 2018) and malathion (Bala et al. 2017) have also been developed. These harmful chemicals lead to groundwater contamination and result in harmful environmental and human health issues.

Toxins primarily derived from bacterial and fungal pathogens are a threat to public health, so their detection and monitoring are important, especially in clinical settings. Aptamers against Gram-positive bacterial toxins such as enterotoxins C1 (Huang et al. 2015) and B (DeGrasse 2012), alpha toxin (Vivekananda et al. 2014), botulinum neurotoxins (Bruno et al. 2012), and anthrax toxins (Kim et al. 2015) have been developed using different aptamer assays to achieve dynamic detecting abilities. Subsequently, lipopolysaccharides (Zhu et al. 2019) and Shiga toxins (Challa et al. 2014) in Gram-negative bacteria were also detected using aptamers, along with aflatoxin B1 (Shim et al. 2014) and ochratoxin A (Ha 2015) in fungal pathogens.

Food-borne pathogens are a threat to agriculture, the food chain, and human health. Since traditional analytical methods are laborious and require specific equipment, there is a need for the quick on-site analysis of large food samples. Food pollutants that can be detected by aptamers include organophosphorus pesticides (Wang et al. 2012), some antibiotics, veterinary drugs, and bacterial toxins (Liu and Zhang 2015). Studies evaluating the efficacy and sensitivity of aptamers to detect food allergens have also been performed (Khedri et al. 2018). Another common antibiotic, sulfamethazine, has been detected in food products via ELAA developed by Le et al. (2018) using a colorimetric aptasensor. This method allowed the detection of the antibiotic at concentrations as low as 0.05 ng mL^{-1} through indirect competitive ELAA, providing a direct and efficient process to detect toxins *in situ*. Aptamer-based detection methods have proven useful to food safety by eliminating lengthy experimental procedures through the direct assessment of food products on site.

Conclusion and Future Perspectives

Aptamer-based technologies have increasingly emerged in recent years due to their vast range of potential applications not only in diagnostics and therapeutics research, but also in other biomedical fields. In particular, the diverse chemical modifications to which aptamers can be subjected along with advances in combinatorial screening methods to select aptamers for virtually any molecular target make them an attractive solution to the ongoing challenges of established immunoassays. The development of ELAA and other aptamer-based point-of-care diagnostic platforms that can detect any type of antigen has provided several advantages over antibody-based assays. These include easier mass production and reusability of MREs, higher assay stability, and longer storage shelf-life, not requiring particularly low-temperature conditions. These advantages also make aptamers more robust MREs, expanding their applicability across a wide variety of biosensors apart from ELISA-based formats. The increasing interest in these aptamer platforms and technologies suggests that a paradigm shift is currently gradually unfolding, with aptamers replacing antibodies as MREs. However, aptamer-based technologies still have a lot of catching up to do, with antibody technologies having several decades head start. Aptamers, having a history of only 20 years, are still in the developmental stage relative to antibodies, which have undergone more than 100 years of development and discovery. More studies are needed to optimize and standardize the methods for screening and characterizing aptamers. The applicability of aptamer-based assays in a wide variety of complex clinical samples should also be taken into consideration. Despite these factors and the technology commercialization bottlenecks encountered, future prospects indicate that aptamers, with their superior properties, will play central roles as recognition molecules for target or biomarker discovery and detection, and drug delivery.

Acknowledgments

The authors thank the Philippine Council for Health Research and Development of the Department of Science and Technology for providing generous funding assistance that enabled the publication of this research material.

References

Abraham, K.M., Roueinfar, M., Ponce, A.T., Lussier, M.E., Benson, D.B., Hong, K.L. 2018. In vitro selection and characterization of a single-stranded DNA aptamer against the herbicide atrazine. *ACS Omega* 3(10): 13576-13583. https://doi.org/10.1021/acsomega.8b01859

Adachi, T., Nakamura, Y. 2019. Aptamers: A review of their chemical properties and modifications for therapeutic application. *Molecules* 24(23): 4229. https://doi.org/10.3390/molecules24234229

Alfavian, H., Gargari, S.L.M., Rasoulinejad, S., Medhat, A. 2017. Development of a DNA aptamer that binds specifically to group A *Streptococcus* serotype M3. *Canadian Journal of Microbiology* 63(2): 160-168. https://doi.org/10.1139/cjm-2016-0495

Asai, R., Nishimura, S.I., Aita, T., Takahashi, K. 2004. In vitro selection of DNA aptamers on chips using a method for generating point mutations. *Analytical Letters* 37(4): 645–656. https://doi.org/10.1081/AL-120029741

Bala, R., Dhingra, S., Kumar, M., Bansal, K., Mittal, S., Sharma, R.K., Wangoo, N. 2017. Detection of organophosphorus pesticide–Malathion in environmental samples using peptide and aptamer based nanoprobes. *Chemical Engineering Journal* 311: 111-116. https://doi.org/10.1016/j.cej.2016.11.070

Balamurugan, S., Obubuafo, A., McCarley, R.L., Soper, S.A., Spivak, D.A. 2008a. Effect of linker structure on surface density of aptamer monolayers and their corresponding protein binding efficiency. *Analytical Chemistry* 80(24): 9630–9634. https://doi.org/10.1021/ac8009559

Balamurugan, S., Obubuafo, A., Soper, S.A., Spivak, D.A. 2008b. Surface immobilization methods for aptamer diagnostic applications. *Analytical and Bioanalytical Chemistry* 390(4): 1009–1021. https://doi.org/10.1007/s00216-007-1587-2

Baldrich, E., Acero, J.L., Reekmans, G., Laureyn, W., O'Sullivan, C.K. 2005a. Displacement enzyme linked aptamer assay. *Analytical Chemistry* 77(15): 4774–4784. https://doi.org/10.1021/ac0502450

Baldrich, E., Restrepo, A., O'Sullivan, C.K. 2005b. Aptasensor development: elucidation of critical parameters for optimal aptamer performance. *Analytical Chemistry* 76(23): 7053–7063. https://doi.org/10.1021/ac049258o

Balogh, Z., Lautner, G., Bardóczy, V., Komorowska, B., Gyurcsanyi, R.E., Meszáros, T. 2010. Selection and versatile application of virus-specific aptamers. *The FASEB Journal* 24(11): 4187–4195. https://doi.org/10.1096/fj.09-144246

Bates, P.J., Laber, D.A., Miller, D.M., Thomas, S.D., Trent, J.O. 2009. Discovery and development of the G-rich oligonucleotide AS1411 as a novel treatment for cancer. *Experimental and Molecular Pathology* 86(3): 151-164. https://doi.org/10.1016/j.yexmp.2009.01.004

Berens, C., Thain, A., Schroeder, R. 2001. A tetracycline-binding RNA aptamer. *Bioorganic & Medicinal Chemistry* 9(10): 2549-2556. https://doi.org/10.1016/S0968-0896(01)00063-3

Borsa, B.A., Tuna, B.G., Hernandez, F.J., Hernandez, L.I., Bayramoglu, G., Arica, M.Y., Ozalp, V.C. 2016. *Staphylococcus aureus* detection in blood samples by silica nanoparticle-oligonucleotides conjugates. *Biosensors and Bioelectronics* 86: 27-32. https://doi.org/10.1016/j.bios.2016.06.023

Bruno, J.G., Richarte, A.M., Carrillo, M.P., Edge, A. 2012. An aptamer beacon responsive to botulinum toxins. *Biosensors and Bioelectronics* 31(1): 240-243. https://doi.org/10.1016/j.bios.2011.10.024

Bruno, J.G., Carrillo, M.P., Phillips, T. 2008. In vitro antibacterial effects of antilipopolysaccharide DNA aptamer-C1qrs complexes. *Folia Microbiologica* 53(4): 295-302. https://doi.org/10.1007/s12223-008-0046-6

Burgstaller, P., Girod, A., Blind, M. 2002. Aptamers as tools for target prioritization and lead identification. *Drug Discovery Today* 7(24): 1221-1228. https://doi.org/10.1016/S1359-6446(02)02522-9

Butler, J.E. 2000. Solid supports in enzyme-linked immunosorbent assay and other solid-phase immunoassays. *Methods* 22(1): 4–23. https://doi.org/10.1006/meth.2000.1031

Centi, S., Tombelli, S., Minunni, M., Mascini, M. 2007. Aptamer-based detection of plasma proteins by an electrochemical assay coupled to magnetic beads. *Analytical Chemistry* 79(4): 1466–1473. https://doi.org/10.1021/ac061879p

Challa, S., Tzipori, S., Sheoran, A. 2014. Selective evolution of ligands by exponential enrichment to identify RNA aptamers against Shiga toxins. *Journal of Nucleic Acids*. https://doi.org/10.1155/2014/214929

Cho, M., Oh, S.S., Nie, J., Stewart, R., Radeke, M.J., Eisenstein, M., Coffey, P.J., Thomson, J.A., Soh, H.T. 2015. Array-based discovery of aptamer pairs. *Analytical Chemistry* 87(1): 821–828. https://doi.org/10.1021/ac504076k

DeGrasse, J.A. 2012. A single-stranded DNA aptamer that selectively binds to *Staphylococcus aureus* enterotoxin B. *PLoS ONE* 7(3): e33410. https://doi.org/10.1371/journal.pone.0033410

Deigan, K.E., FerrÉ-D'AmarÉ, A.R. 2011. Riboswitches: discovery of drugs that target bacterial gene-regulatory RNAs. *Accounts of Chemical Research* 44(12): 1329–1338. https://doi.org/10.1021/ar200039b

Drolet, D.W., Moon-McDermott, L., Romig, T.S. 1996. An enzyme-linked oligonucleotide assay. *Nature Biotechnology* 14(8): 1021–1025. https://doi.org/10.1038/nbt0896-1021

Ellington, A.D., Szostak, J.W. 1990. In vitro selection of RNA molecules that bind specific ligands. *Nature* 346(6287): 818–822. https://doi.org/10.1038/346818a0

Espiritu, C.A.L., Justo, C.A.C., Rubio, M.J., Svobodova, M., Bashammakh, A.S., Alyoubi, A.O., Rivera, W.L., Rollon, A.P., O'Sullivan, C.K. 2018. Aptamer selection against a *Trichomonas vaginalis* adhesion protein for diagnostic applications. *ACS Infectious Diseases* 4(9): 1306-1315. https://doi.org/10.1021/acsinfecdis.8b00065

Farzin, L., Shamsipur, M., Sheibani, S. 2017. A review: Aptamer-based analytical strategies using the nanomaterials for environmental and human monitoring of toxic heavy metals. *Talanta* 174: 619-627. https://doi.org/10.1016/j.talanta.2017.06.066

Ferrara, N. 2004. Vascular endothelial growth factor: basic science and clinical progress *Endocrine Reviews* 25(4): 581-611. https://doi.org/10.1210/er.2003-0027

Ghavamipour, F., Rahmani, H., Shanehsaz, M., Khajeh, K., Mirshahi, M., Sajedi, R.H. 2020. Enhanced sensitivity of VEGF detection using catalase-mediated chemiluminescence immunoassay based on CdTe QD/H2O2 system. *Journal of Nanobiotechnology* 18(1): 93. https://doi.org/10.1186/s12951-020-00648-9

Go, G., Lee, C-S., Yoon, Y.M., Lim, J.H., Kim, T.H., Lee, S.H. 2021. PrP^C aptamer conjugated–gold nanoparticles for targeted delivery of doxorubicin to colorectal cancer cells. *International Journal of*

Molecular Sciences 22(4): 1976. https://doi.org/10.3390/ijms22041976

Gong, X., Li, X., Qing, T., Zhang, P., Feng, B. 2019. Amplified colorimetric detection of tetracycline based on an enzyme-linked aptamer assay with multivalent HRP-mimicking DNAzyme. *Analyst* 144(6): 1948-1954. https://doi.org/10.1039/C8AN02284D

Gordon, E.M., Barrett, R.W., Dower, W.J., Fodor, S.P.A., Gallop. M.A. 1994. Applications of combinatorial technologies to drug discovery. 2. Combinatorial organic synthesis, library screening strategies, and future directions. *Journal of Medicinal Chemistry* 37(10): 1385–1401. https://doi.org/10.1021/jm00036a001

Ha, T.H. 2015. Recent advances for the detection of ochratoxin A. *Toxins* 7(12): 5276-5300. https://doi.org/10.3390/toxins7124882

Hasegawa, H., Savory, N., Abe, K., Ikebukuro, K. 2016. Methods for improving aptamer binding affinity. *Molecules* 21(4): 421. https://doi.org/10.3390/molecules21040421

Healy, J.M., Lewis, S.D., Kurz, M., Boomer, R.M., Thompson, K.M., Wilson, C., McCauley, T.G. 2004. Pharmacokinetics and biodistribution of novel aptamer compositions. *Pharmaceutical Research* 21(12): 2234–2246. https://doi.org/10.1007/s11095-004-7676-4

Hill, A.V. 1910. The possible effects of the aggregation of the molecules of haemoglobin on its dissociation curves. *The Journal of Physiology* 40: 4-7.

Hong, X., Yan, H., Xie, F., Wang, K., Wang, Q., Huang, H., Yang, K., et al. 2019. Development of a novel ssDNA aptamer targeting neutrophil gelatinase-associated lipocalin and its application in clinical trials. *Journal of Translational Medicine* 17(1): 1-14. https://doi.org/10.1186/s12967-019-1955-7

Huang, Y., Chen, X., Duan, N., Wu, S., Wang, Z., Wei, X., Wang, Y. 2015. Selection and characterization of DNA aptamers against *Staphylococcus aureus* enterotoxin C1. *Food Chemistry* 166: 623-629. https://doi.org/10.1016/j.foodchem.2014.06.039

Jarczewska, M., Malinowska, E. 2020. The application of antibody-aptamer hybrid biosensors in clinical diagnostics and environmental analysis. *Analytical Methods* 12(25): 3183–3199. https://doi.org/10.1039/d0ay00678e

Jeong, S., Han, S.R., Lee, Y.J., Kim, J.H., Lee, S-W. 2010. Identification of RNA aptamer specific to mutant KRAS protein. *Oligonucleotides* 20(3): 155–161. https://doi.org/10.1089/oli.2010.0231

Kaur, H. 2018. Recent developments in cell-SELEX technology for aptamer selection. *Biochimica et Biophysica Acta (BBA)-General Subjects* 1862(10): 2323-2329. https://doi.org/10.1016/j.bbagen.2018.07.029

Kaur, H., Yung, L-Y.L. 2012. Probing high affinity sequences of DNA aptamer against $VEGF_{165}$. *PLoS ONE* 7(2): e31196. https://doi.org/10.1371/journal.pone.0031196

Khedri, M., Ramezani, M., Rafatpanah, H., Abnous, K. 2018. Detection of food-born allergens with aptamer-based biosensors. *TrAC Trends in Analytical Chemistry* 103: 126-136. https://doi.org/10.1016/j.trac.2018.04.001

Kim, J., Gedi, V., Lee, S-C., Cho, J-H., Moon, J-Y., Yoon, M-Y. 2015. Advances in anthrax detection: overview of bioprobes and biosensors." *Applied Biochemistry and Biotechnology* 176(4): 957-977. https://doi.org/10.1007/s12010-015-1625-z

Kratschmer, C., Levy, M. 2017. Effect of chemical modifications on aptamer stability in serum. *Nucleic Acid Therapeutics* 27(6): 335–344. https://doi.org/10.1089/nat.2017.0680

Kulabhusan, P.K., Hussain, B., Yüce, M. 2020. Current perspectives on aptamers as diagnostic tools and therapeutic agents. *Pharmaceutics* 12(7): 646. https://doi.org/10.3390/pharmaceutics12070646

Kuwahara, M., Sugimoto, N. 2010. Molecular evolution of functional nucleic acids with chemical modifications. *Molecules* 15(8): 5423–5444. https://doi.org/10.3390/molecules15085423

Lakhin, A.V., Tarantul, V.Z., Gening, L.V. 2013. Aptamers: problems, solutions and prospects. *Acta Naturae* 5(4): 34–43.

Lamont, E.A., He, L., Warriner, K., Labuza, T.P., Sreevatsan, S. 2011. A single DNA aptamer functions as a biosensor for ricin. *Analyst* 136(19): 3884–3895. https://doi.org/10.1039/c1an15352h

Le, T., Sun, Q., Xie, Y., Shu, L., Liu, J., Xu, J., Xiong, J., Cao, X. 2018. A highly sensitive aptasensor for sulfamethazine detection using an enzyme-linked aptamer assay. *Food Analytical Methods* 11(10): 2778-2787. https://doi.org/10.1007/s12161-018-1258-2

Lee, K.H., Zeng, H. 2017. Aptamer-based ELISA assay for highly specific and sensitive detection of Zika NS1 protein. *Analytical Chemistry* 89(23): 12743-12748. https://doi.org/10.1021/acs.analchem.7b02862

Liu, X., Zhang, X. 2015. Aptamer-based technology for food analysis. *Applied Biochemistry and Biotechnology* 175: 603–624 (2015). https://doi.org/10.1007/s12010-014-1289-0

Ma, H., Ó'Fágáin, C., O'Kennedy, R. 2020. Antibody stability: a key to performance - analysis, influences and improvement." *Biochimie* 177: 213–225. https://doi.org/10.1016/j.biochi.2020.08.019

Ma, X., Jiang, Y., Jia, F., Yu, Y., Chen, J., Wang, Z. 2014. An aptamer-based electrochemical biosensor for the detection of *Salmonella*. *Journal of Microbiological Methods* 98: 94-98. https://doi.org/10.1016/j.mimet.2014.01.003

Mahurpawar, M. 2015. Effects of heavy metals on human health. *International Journal of Research-Granthaalayah* 3(9SE):1-7. https://doi.org/10.29121/granthaalayah.v3.i9SE.2015.3282

Mao, K., Zhang, H., Wang, Z., Cao, H., Zhang, K., Li, X., Yang, Z. 2020. Nanomaterial-based aptamer sensors for arsenic detection. *Biosensors and Bioelectronics* 148: 111785. https://doi.org/10.1016/j.bios.2019.111785

Marnissi, B., Kamali-Moghaddam, M., Ghram, A., Hmila, I. 2020. Generation of ssDNA aptamers as diagnostic tool for Newcastle avian virus. *PLoS ONE* 15(8): e0237253. https://doi.org/10.1371/journal.pone.0237253

Marton, S., Cleto, F., Krieger, M.A., Cardoso, J. 2016. Isolation of an aptamer that binds specifically to *E. coli*. *PloS ONE* 11(4): e0153637. https://doi.org/10.1371/journal.pone.0153637

Mehta, J., Dorst, B.V., Rouah-Martin, E., Herrebout, W., Scippo, M-L., Blust, R., Robbens, J. 2011. In vitro selection and characterization of DNA aptamers recognizing chloramphenicol. *Journal of Biotechnology* 155(4): 361-369. https://doi.org/10.1016/j.jbiotec.2011.06.043

Mutreja R., Baba S.A., Navani N.K. 2019. Nucleic acid aptamers as emerging tools for diagnostics and theranostics. In: Batra J., Srinivasan S. (eds) *Theranostics. Methods in Molecular Biology*, vol 2054. Humana, New York, NY. https://doi.org/10.1007/978-1-4939-9769-5_14

Nazari, M., Gargari, S.L.M., Lotfi, A.S., Rassaee, M.J., Taheri, R.A. 2019. Aptamer-based sandwich assay for measurement of thymidine kinase 1 in serum of cancerous patients. *Biochemistry* 58(18): 2373-2383. https://doi.org/10.1021/acs.biochem.8b01284

Nezlin, R. 2016. Use of aptamers in immunoassays. *Molecular Immunology* 70: 149–154. https://doi.org/10.1016/j.molimm.2015.12.009

Ng, E.W.M., Shima, D.T., Calias, P., Cunningham, E.T., Guyer, D.R., Adamis, A.P. 2006. Pegaptanib, a targeted anti-VEGF aptamer for ocular vascular disease. *Nature Reviews Drug Discovery* 5(2): 123–132. https://doi.org/10.1038/nrd1955

Ospina-Villa, J.D., Zamorano-Carrillo, A., Castañón-Sánchez, C.A., Ramírez-Moreno, E., Marchat, L.A. 2016. Aptamers as a promising approach for the control of parasitic diseases. *Brazilian Journal of Infectious Diseases* 20: 610-618. https://doi.org/10.1016/j.bjid.2016.08.011

Parekh, P., Tang, Z., Turner, P.C., Moyer, R.W., Tan, W. 2010. Aptamers recognizing glycosylated hemagglutinin expressed on the surface of vaccinia virus-infected cells. *Analytical Chemistry* 82(20): 8642-8649. https://doi.org/10.1021/ac101801j

Park, H., Paeng, I.R. 2011. Development of direct competitive enzyme-linked aptamer assay for determination of dopamine in serum. *Analytica Chimica Acta* 685(1): 65-73. https://doi.org/10.1016/j.aca.2010.11.010

Pleshakova, T.O., Kaysheva, A.L., Bayzyanova, J.M., Anashkina, A.S., Uchaikin, V.F., Ziborov, V.S., Konev, V.A., Archakov, A.I., Ivanov, Y.D. 2018. The detection of hepatitis C virus core antigen using afm chips with immobolized aptamers. *Journal of Virological Methods* 251: 99-105. https://doi.org/10.1016/j.jviromet.2017.10.015

Qi, Y., Ma, J., Chen, X., Xiu, F-R., Chen, Y., Lu, Y. 2020. Practical aptamer-based assay of heavy metal mercury ion in contaminated environmental samples: convenience and sensitivity. *Analytical and Bioanalytical Chemistry* 412(2): 439-448. https://doi.org/10.1007/s00216-019-02253-8

Rosenberg, J.E., Bambury, R.M., Van Allen, E.M., Drabkin, H.A., Lara, P.N., Harzstark, A.L., Wagle, N., et al. 2014. A phase II trial of AS1411 (a novel nucleolin-targeted DNA aptamer) in metastatic renal cell carcinoma. *Investigational New Drugs* 32(1): 178-187. https://doi.org/10.1007/s10637-013-0045-6

Rusconi, C.P., Roberts, J.D., Pitoc, G.A., Nimjee, S.M., White, R.R., Quick Jr, G., Scardino, E., Fay, W.P., Sullenger, B.A. 2004. Antidote-mediated control of an anticoagulant aptamer in vivo. *Nature Biotechnology* 22(11): 1423–1428. https://doi.org/10.1038/nbt1023

Schlesinger, S.R., Lahousse, M.J., Foster, T.O., Kim, S-K. 2011. Metallo-β-lactamases and aptamer-based inhibition. *Pharmaceuticals* 4(2): 419-428. https://doi.org/10.3390/ph4020419

Sela-Culang, I., Kunik, V., Ofran, Y. 2013. The structural basis of antibody-antigen recognition. *Frontiers in Immunology* 4: 302. https://doi.org/10.3389/fimmu.2013.00302

Shangguan, J., Li, Y. He, D., He, X., Wang, K., Zou, Z., Shi, H. 2015. A combination of positive dielectrophoresis driven on-line enrichment and aptamer-fluorescent silica nanoparticle label for rapid and sensitive detection of *Staphylococcus aureus*. *Analyst* 140(13): 4489-4497. https://doi.org/10.1039/C5AN00535C

Shim, W-B., Kim, M.J., Mun, H., Kim, M-G. 2014. An aptamer-based dipstick assay for the rapid and simple detection of aflatoxin B1. *Biosensors and Bioelectronics* 62: 288-294. https://doi.org/10.1016/j.bios.2014.06.059

Shukla, A.A., Wolfe, L.S., Mostafa, S.S., Norman, C. 2017. Evolving trends in mAb production processes. *Bioengineering & Translational Medicine* 2(1): 58–69. https://doi.org/10.1002/btm2.10061

Sinha, K., Mukhopadhyay, C.D. 2020. Quantitative detection of neurotransmitter using aptamer: From diagnosis to therapeutics. *Journal of Biosciences* 45(1): 1-12. https://doi.org/10.1007/s12038-020-0017-x

Sola, M., Menon, A.P., Moreno, B., Meraviglia-Crivelli, D., Soldevilla, M.M., Cartón-García, F., Pastor, F. 2020. "Aptamers against live targets: is in vivo SELEX finally coming to the edge? *Molecular Therapy - Nucleic Acids* 21: 192–204. https://doi.org/10.1016/j.omtn.2020.05.025

Song, K-M., Lee, S., Ban, C. 2012. Aptamers and their biological applications. *Sensors (Basel)* 12(1): 612–631. https://doi.org/10.3390/s120100612

Song, M.Y., Nguyen, D., Hong, S.W., Kim, B.C. 2017. Broadly reactive aptamers targeting bacteria belonging to different genera using a sequential toggle cell-SELEX. *Scientific Reports* 7:43641. https://doi.org/10.1038/srep43641

Soundararajan, S., Chen, W., Spicer, E.K., Courtenay-Luck, N., Fernandes, D.J. 2008. The nucleolin targeting aptamer AS1411 destabilizes Bcl-2 messenger RNA in human breast cancer cells. *Cancer Research* 68(7): 2358-2365. https://doi.org/10.1158/0008-5472.CAN-07-5723

Suh, S-K., Song, S., Oh, H-B., Hwang, S-H., Hah, S.S. 2014. Aptamer-based competitive binding assay for one-step quantitation of hepatitis B surface antigen. *Analyst* 139(17): 4310-4314. https://doi.org/10.1039/C4AN00619D

Sullenger, B.A., Gallardo, H.F., Ungers, G.E., Gilboa, E. 1990. Overexpression of TAR sequences renders cells resistant to human immunodeficiency virus replication. *Cell* 63(3): 601–608. https://doi.org/10.1016/0092-8674(90)90455-n

Sullenger, B.A., Gallardo, H.F., Ungers, G.E., Gilboa, E. 1991. Analysis of trans-acting response decoy RNA-mediated inhibition of human immunodeficiency virus type 1 transactivation. *Journal of Virology* 65(12): 6811–6816. https://doi.org/10.1128/JVI.65.12.6811-6816. 1991

Toh, S.Y., Citartan, M., Gopinath, S.C.B., Tang, T-H. 2015. Aptamers as a replacement for antibodies in enzyme-linked immunosorbent assay. *Biosensors and Bioelectronics* 64: 392–403. https://doi.org/10.1016/ j.bios.2014.09.026

Tombelli, S., Minunni, M., Luzi, E., Mascini, M. 2005. Aptamer-based biosensors for the detection of HIV-1 Tat protein. *Bioelectrochemistry* 67(2): 135-141. https://doi.org/10.1016/j.bio elechem. 2004.04.011

Tuerk, C, Gold, L. 1990. Systematic evolution of ligands by exponential enrichment: RNA ligands to bacteriophage T4 DNA polymerase. *Science* 249(4968): 505-510. https://doi.org/10.1126/science.220 0121.

Urak, K.T., Shore, S., Rockey, W.M., Chen, S-J., McCaffrey, A.P., Giangrande, P.H. 2016. In vitro RNA SELEX for the generation of chemically-optimized therapeutic RNA drugs. *Methods* 103: 167–174. https://doi.org/10.1016/j.ymeth.2016.03.003

Urmann, K., Modrejewski, J., Scheper, T., Walter, J.G. 2017. Aptamer-modified nanomaterials: principles and applications. *Bio Nano Materials* 18(1–2). https://doi.org/10.1515/bnm-2016-0012

van Dijk, M., van Dijk, A.D.J., Hsu, V., Rolf, B., Bonvin, A.M.M.J. 2006. Information-driven protein-DNA docking using HADDOCK: it is a matter of flexibility. *Nucleic Acids Research* 34(11): 3317–3325. https://doi.org/10.1093/nar/gkl412

Vargas-Montes, M., Cardona, N., Moncada, D.M., Molina, D.A., Zhang, Y., Gómez-Marín, J.E. 2019. Enzyme-linked aptamer assay (ELAA) for detection of *Toxoplasma* ROP18 protein in human serum. *Frontiers in Cellular and Infection Microbiology* 9:386. https://doi.org/10.3389/fcimb.2019.00386

Vinores, S.A. 2006. Pegaptanib in the treatment of wet, age-related macular degeneration. *International Journal of Nanomedicine* 1(3): 263-268. https://doi.org/10.1016/j.exer.2006.02.010

Vivekananda, J., Kiel, J.L. 2006. Anti-*Francisella tularensis* DNA aptamers detect tularemia antigen from different subspecies by aptamer-linked immobilized sorbent assay. *Laboratory Investigation* 86(6): 610–618. https://doi.org/10.1038/labinvest.3700417

Vivekananda, J., Salgado, C., Millenbaugh, N.J. 2014. DNA aptamers as a novel approach to neutralize *Staphylococcus aureus* α-toxin. *Biochemical and Biophysical Research Communications* 444(3): 433-438. https://doi.org/10.1016/j.bbrc.2014.01.076

Vorobyeva, M.A., Davydova, A.S., Vorobjev, P.E., Venyaminova, A.G. 2018. Key aspects of nucleic acid library design for in vitro selection. *International Journal of Molecular Sciences* 19(2): 470. https://doi.org/10.3390/ijms19020470

Wang, L., Liu, X., Zhang, Q., Zhang, C., Liu, Y., Tu, K., Tu, J. 2012. Selection of DNA aptamers that bind to four organophosphorus pesticides. *Biotechnology Letters* 34(5): 869-874. https://doi.org/10.1007/s10529-012-0850-6

Wang, R., Zhao, J., Jiang, T., Kwon, Y.M., Lu, H., Jiao, P., Liao, M., Li, Y. 2013. Selection and characterization of DNA aptamers for use in detection of avian influenza virus H5N1." *Journal of Virological Methods* 189(2): 362-369. https://doi.org/10.1016/j.jviromet.2013.03.006

Welch, N.G., Scoble, J.A., Muir, B.W., Pigram, P.J. 2017. Orientation and characterization of immobilized antibodies for improved immunoassays (Review). *Biointerphases* 12(2): 02D301. https://doi.org/10.1116/1.4978435

Wildner, S., Huber, S., Regl, C., Huber, C.G., Lohrig, U., Gadermaier, G. 2019. Aptamers as quality control tool for production, storage and biosimilarity of the anti-CD20 biopharmaceutical rituximab."

Scientific Reports 9(1): 114. https://doi.org/10.1038/s41598-018-37624-1

Wongphatcharachai, M., Wang, P., Enomoto, S., Webby, R.J., Gramer, M.R., Amonsin, A., Sreevatsan, S. 2013. Neutralizing DNA aptamers against swine influenza H3N2 viruses. *Journal of Clinical Microbiology* 51(1): 46–54. https://doi.org/10.1128/JCM.02118-12

Wu, J., Zhu, Y., Xue, F., Mei, Z., Yao, L., Wang, X., Zheng, L. et al. 2014. Recent trends in SELEX technique and its application to food safety monitoring. *Mikrochimica Acta* 181(5–6): 479–491. https://doi.org/10.1007/s00604-013-1156-7

Wu, Y., Zhan, S., Wang, L., Zhou, P. 2014. Selection of a DNA aptamer for cadmium detection based on cationic polymer mediated aggregation of gold nanoparticles. *Analyst* 139(6): 1550-1561. https://doi.org/10.1039/C3AN02117C

Yoshida, W., Mochizuki, E., Takase, M., Hasegawa, H., Morita, Y., Yamazaki, H., Sode, K., Ikebukuro, K. 2009. Selection of DNA aptamers against insulin and construction of an aptameric enzyme subunit for insulin sensing. *Biosensors & Bioelectronics* 24(5): 1116–1120. https://doi.org/10.1016/j.bios.2008.06.016

Young, D.D., Deiters, A. 2008. Light-regulated RNA-small molecule interactions. *Chem Bio Chem* 9(8): 1225–1228. https://doi.org/10.1002/cbic.200800051

Yu, B., Tai, H.C., Xue, W., Lee, L.J., Lee, R.J. 2010. Receptor-targeted nanocarriers for therapeutic delivery to cancer. *Molecular Membrane Biology* 27(7): 286–298. https://doi.org/10.3109/09687688.2010.521200

Yuan, J., Wu, S., Duan, N., Ma, X., Xia, Y., Chen, J., Ding, Z., Wang, Z. 2014. A sensitive gold nanoparticle-based colorimetric aptasensor for *Staphylococcus aureus*. *Talanta* 127:163-168. https://doi.org/10.1016/j.talanta.2014.04.013

Zemora, G., Waldsich, C. 2010. RNA folding in living cells. *RNA Biology* 7(6): 634–641. https://doi.org/10.4161/rna.7.6.13554

Zhang, L., Mu, C., Zhang, T., Wang, Y., Wang, Y., Fan, L., Liu, C. et al. 2020. Systemic delivery of aptamer-conjugated XBP1 siRNA nanoparticles for efficient suppression of HER2+ breast cancer. *ACS Applied Materials & Interfaces* 12(29): 32360-32371. https://doi.org/10.1021/acsami.0c07353

Zhang, Y., Lai, B.S., Juhas, M. 2019. Recent advances in aptamer discovery and applications. *Molecules* 24(5): 941. https://doi.org/10.3390/molecules24050941

Zhou, J., Rossi, J. 2017. Aptamers as targeted therapeutics: current potential and challenges. *Nature Reviews Drug Discovery* 16(3): 181–202. https://doi.org/10.1038/nrd.2016.199.

Zhu, L., Li, S., Shao, X., Feng, Y., Xie, P., Luo, Y., Huang, K., Xu, W. 2019. Colorimetric detection and typing of *E. coli* lipopolysaccharides based on a dual aptamer-functionalized gold nanoparticle probe. *Microchimica Acta* 186(2): 111. https://doi.org/10.1007/s00604-018-3212-9

Zou, X., Wu, J., Gu, J., Shen, L., Mao, L. 2019. Application of aptamers in virus detection and antiviral therapy. *Frontiers in Microbiology* 10:1462. https://doi.org/10.3389/fmicb.2019.01462

Editor's Contact Information

Jose Ma. M. Angeles, PhD
Editor and Associate Professor
Department of Parasitology
College of Public Health
University of the Philippines Manila, Manila, Philippines
jmangeles3@up.edu.ph

INDEX

#

A

B

C

O

P

Q

R

S

T

U

V